# Principles and Practices of Electrical Machines

NIPA® GENX ELECTRONIC RESOURCES & SOLUTIONS P. LTD.
New Delhi-110 034

## About the Authors

**Naveen Jain** received his B. Eng. degree in Electrical Engineering and his M. Eng. degree in Power Systems from Malaviya National Institute of Technology Jaipur (Erstwhile MREC Jaipur), India, in 1997 and 1999, respectively. He received his Ph.D. degree in Electrical Engineering from the Indian Institute of Technology Kanpur, Kanpur, India, in 2013. He has over 20 years of teaching and research experience. Presently, he is working as an Associate Professor in the Department of Electrical Engineering at College of Technology and Engineering, Udaipur, India. He has published many research papers in reputed journals and National/International conferences.

He is a reviewer for several international journals and international conferences. He is a Fellow of the Institution of Electronics and Telecommunication Engineers (India), Senior Member of Institute of Electrical and Electronics Engineers (USA), Life Member of Indian Society for Technical Education, and Life Member of Institution of Engineers (India). His research interests include planning of distributed generations, renewable energy grid integration, and application of modern optimization methods in power systems, and technical issues in electricity markets.

**Umesh Agarwal** graduated in Electrical Engineering from the College of Technology and Engineering (CTAE, Udaipur) India, in 2012. He received his M. Tech. degree in Power Systems in 2015 from Rajasthan Technical University (RTU, Kota), India. Further, he has completed his Doctoral degree (PhD) from College of Technology and Engineering (CTAE, Udaipur) India, in 2022. Currently, he is working as an Associate Professor at JIET, Jodhpur. He has over 08 years of teaching and research experience.

He has published many research papers in reputed journals and National/ International conferences. He is reviewer of many reputed journals and conferences. He is a member of Soft Computing Research Society (SCRS). His areas of interest are power system restructuring, AI application to the power system, reliability analysis, and planning of distributed energy.

# Principles and Practices of Electrical Machines

**Naveen Jain**

Associate Professor
Department of Electrical Engineering
College of Technology and Engineering
Udaipur, Rajasthan, India

**Umesh Agarwal**

Associate Professor
Jodhpur Institute of Engineering and Technology (JIET)
Jodhpur, Rajasthan, India

**NIPA® GENX ELECTRONIC RESOURCES & SOLUTIONS P. LTD.**

New Delhi-110 034

**NIPA® GENX ELECTRONIC RESOURCES & SOLUTIONS P. LTD.**

101,103, Vikas Surya Plaza, CU Block
L.S.C.Market, Pitam Pura, New Delhi-110 034
Ph : +91 11 27341616, 27341717, 27341718
E-mail:newindiapublishingagency@gmail.com
www: www.nipabooks.com

***For customer assistance, please contact***
Phone: + 91-11-27 34 17 17
Fax: + 91-11- 27 34 16 16



ISBN: 978-93-58874-01-3



Composed and Designed by NIPA®.

# Preface

This book has been written preliminary as a reference for electrical machine practical course. Practical's are often conducted with a dull monotony, place components, hook up wires, connect wires and record results that are expected well before hand. To add misery, there is a great paucity of books and manuals that can motivate and guide students during experimentation.

But experimentation is fun. The student armed as he is with theory, often realizes that he has traversed into an unknown realm where there are surprise results, derivations and hurdles one could not have envisaged in theory. And also, a realm where, with some patience, some intelligent discussion, and some guidance, one can also find a solution, an alternative betterment in terms of design, and one can learn in the process. This book intends to remove the present gap in the field of practical experimentation and provide the necessary motivation for performing the experiments with a constructive attitude.

The experiments in this book cover a very important range of topics supporting a complementary theoretical course on electrical machines. The objectives that are to be achieved at the end of the experimentation are stated at the beginning to provide a clear direction and goal to the students. The question bank is designed to analysis the result, correlate theoretical knowledge with observation and understands the practical limitations. Many of the questions make the student probe further in to the topic to stir his interest. Each experiment also has a section aimed at concretizing the knowledge gained by the experiment and providing a nutshell understanding of the theory involved. Instruments required to perform the experiment are numbered. Step by step procedure, along with complete tables and circuit diagram are detailed.

The range of work covered in the electrical machine course is now so wide that it is almost impracticable to have a manual that could provide all experimentations on this branch. The manual is thus prepared selecting more fundamental experiments that would provide a total coverage to the chapters with transformer, D.C. machine, Induction machines and Synchronous machines. Though, the experiments given in this book are not exhaustive, the authors timed to put the experiments in such fashion so that they are equipped with brief theoretical details, various stages of the experiments and related questions. The section on theoretical background gives a clear and concise understanding of the theory involved without elaborate proofs, which can be found in textbooks.

It is hoped that this effort will ultimately help students to sharpen their power to observation, enhance their analysis and design skills and make the process of experimentation an enjoyable experience of learning.

**Authors**

# Acknowledgement

We express our heartfelt thanks to Dr. P. K. Singh, Dean, College of Technology and Engineering, Maharana Pratap University of Agriculture and Technology Udaipur, who has been a source of inspiration and whose co-operation helped us to complete this work.

We thankfully acknowledge the contribution of various authors of different books, magazines, data manuals, journals, websites etc. from where materials have been collected to enrich the content of the manual.

We must thank faculty members, staff, technical assistants, technicians and Electrical Engineering Department without whose dedication; compilation of this work would have not been possible.

Specially, we would like to thank Mr. Vahid Hussain, Ms. Bhumika Singh Rathore and Mr. Hukmichand for their valuable contribution for successfully completion of the work.

Finally, I would again like to thank the entire faculty in the Department and those people who directly or indirectly helped in successful completion of this work.

We sincerely welcome the comments and suggestions from the readers for any kind of improvement in the book, which may be communicated

**Authors**

# Contents

# Laboratory Practices

## Safety Rules

1. Safety is of paramount importance in the electrical engineering laboratories.
2. Electricity never excuses careless persons. So, exercise enough care and attention in handling electrical equipment and follow safety practices in the laboratory.
3. Avoid direct contact with any voltage source and power line voltages. (Otherwise, any such contact may subject you to electrical shock).
4. Wear rubber-soled shoes. (To insulate you from earth so that even if you accidentally contact a live point, current will not flow through your body to earth and hence you will be protected from electrical shock).
5. Wear laboratory-coat and avoid loose clothing. (Loose clothing may get caught on an equipment/instrument and this may lead to an accident particularly if the equipment happens to be a rotating machine).
6. Girl students should have their hair tucked under their coat or have it in a knot.
7. Do not wear any metallic rings, bangles, bracelets, wristwatches and neck chains. (When you move your hand/body, such conducting items may create a short circuit or may touch a live point and thereby subject you to electrical shock).
8. Be certain that your hands are dry and that you are not standing on wet floor. (Wet parts of the body reduce the contact resistance thereby increasing the severity of the shock).
9. Ensure that the power is off before you start connecting up the circuit. (otherwise you will be touching the live parts in the circuit).
10. Get your circuit diagram approved by the staff member and connect up the circuit strictly as per the approved circuit diagram.

11. Check power chords for any sign of damage and be certain that the chords use safety plugs and do not defeat the safety feature of these plugs by using ungrounded plugs.
12. When using connection leads, check for any insulation damage in the leads and avoid such defective leads.
13. Do not defeat any safety devices such as fuse or circuit breaker by shorting across it. Safety devices protect you and your equipment.
14. Switch on the power to your circuit and equipment only after getting them checked up and approved by the staff member.
15. Take the measurement with one hand in your pocket. (To avoid shock in case you accidentally touch two points at different potentials with your two hands).
16. Do not make any change in the connection without the approval of the staff member.
17. In case you notice any abnormal condition in your circuit (like insulation heating up, resistor heating up etc), switch off the power to your circuit immediately and inform the staff member.
18. Keep hot soldering iron in the holder when not in use.
19. After completing the experiment show your readings to the staff member and switch off the power to your circuit after getting approval from the staff member.
20. While performing load-tests in the electrical machines' laboratory using the brake-drums:

    i. Avoid the brake-drum from getting too hot by putting just enough water into the brake-drum at intervals; use the plastic bottle with a nozzle (available in the laboratory) to pour the water.(when the drum gets too hot, it will burn out the braking belts)

    ii. Do not stand in front of the brake-drum when the supply to the load-test circuit is switched off. (Otherwise, the hot water in the brake-drum will splash out on you)

    iii. After completing the load-test, suck out the water in the brake-drum using the plastic bottle with nozzle and then dry off the drum with a sponge which is available in the laboratory (the water, if allowed to remain in the brake-drum, will corrode it).

21. Determine the correct rating of the fuse/s to be connected in the circuit after understanding correctly the type of the experiment to be performed: no-load test or full-load test, the maximum current expected in the circuit and accordingly use that fuse-rating. (While an over-rated fuse will damage the equipment and other instruments like ammeters and watt-meters in case of over load, an under-rated fuse may not allow one even to start the experiment).

22. At the time of starting a motor, the ammeter connected in the armature circuit overshoots, as the starting current is around 5 times the full load rating of the motor. Moving coil ammeters being very delicate may get damaged due to high starting current. A switch has been provided on such meters to disconnect the moving coil of the meter during starting. This switch should be closed after the motor attains full speed. Moving iron ammeters and current coils of watt meters are not so delicate and hence these can stand short time overload due to high starting current. No such switch is therefore provided on these meters. Moving iron meters are cheaper and more rugged compared to moving coil meters. Moving iron meters can be used for both a.c. And d.c. Measurement. Moving coil instruments are however more sensitive and more accurate as compared to their moving iron counterparts and these can be used for d.c. Measurements only. Good features of moving coil instruments are not of much consequence for you as other sources of errors in the experiments are many times more than those caused by these meters.

23. Some students have been found to damage meters by mishandling in the following ways:

    i. Keeping unnecessary material like books, lab records, unused meters etc. Causing meters to fall down the table.

    ii. Putting pressure on the meter (especially glass) while making connections or while talking or listing somebody.

**Copy these rules in your Lab Record.**

**Observe this yourself and help your friends to observe.**

I have read and understand these rules and procedures. I agree to abide by these rules and procedures at all times while using these facilities. I understand that failure to follow these rules and procedures will result in my immediate dismissal from the laboratory and additional disciplinary action may be taken.

## Guidelines for Preparing Laboratory Notebook

The laboratory notebook is a record of all work pertaining to the experiment. This record should be sufficiently complete so that you or anyone else of similar technical background can duplicate the experiment and data by simply following your laboratory notebook. Record everything directly into the notebook during the experiment. Do not use scratch paper for recording data. Do not trust your memory to fill in the details at a later time.

Organization in your notebook is important. Descriptive headings should be used to separate and identify the various parts of the experiment. Record data in chronological order. A neat, organized and complete record of an experiment is just as important as the experimental work.

1. **Heading**: The experiment identification (number) should be at the top of first page. Your name and date should be mentioned at the top of each page of days experimental.
2. **Object:** A brief complete statement of what you intend to find out or verify in the experiment should be at the beginning of each experiment.
3. **Equipment List:** List those items of equipment which have a direct effect on the accuracy of the data. It may be necessary later to locate specific items of equipment for rechecks if discrepancies develop in the results.
4. **Circuit Diagram:** A circuit diagram should be drawn and labelled so that the actual experiment circuitry could be easily duplicated at any time in the future. Be especially careful to record all circuit changes made during the experiment.
5. **Theory:** As per experiment.
6. **Procedure:** In general, lengthy explanations of procedures are unnecessary so be in brief, short commentaries alongside the corresponding data may be used. Keep in mind the fact that the experiment must be reproducible from the information given in your notebook.
7. **Data:** Think carefully about what data is required and prepare suitable data tables. Record instruments reading directly. Do not use calculated results in place of direct data however; calculated results may be recorded in the same table with the direct data. Data tables should be clearly identified and each data column labelled and headed by the proper units of measure.
8. **Calculations:** Not always necessary but equations and sample calculations are often given to illustrate the treatment of the experimental data in obtaining the results.

9. **Results:** The results should be presented in a form which makes the interpretation easy. Large amounts of numerical results are generally presented in graphical form. Tables are generally used for small amounts of results. Theoretical and experimental results should be on the same graph or arrange in the same table in a way for easy correlation of these results.

10. **Graphs:** Graphs are used to present large amounts of data in a concise visual form. Data to be presented in graphical form should be plotted in the laboratory so that any questionable data points can be checked while the experiment is still set up. The grid lines in the notebook can be used for most graphs. If special graph paper is required, affix the graph permanently into the notebook. Give all graphs a short descriptive title. Label and scale the axes. Use units of measure. Label each curve if more than one on a graph.

11. **Conclusion:** This is your interpretation of the results of the experiment as an engineer. Be brief and specific. Give reasons for important discrepancies.

12. **Precautions**

## Troubleshooting Hints

1 Be Sure that the power is turned ON.

2 Be sure the ground connections are common.

3 Be sure the circuit you build is identical to your circuit diagram (Do a node-by-node check).

4 Be sure that the supply voltages are correct.

5 Be sure that the equipment is set up correctly and you are measuring the correct parameters.

6 If steps 1 to 5 are correct then you probably have used a component with the wrong value or one that doesn't work. It is also possible that the equipment does not work (although this is not probable or the protoboard you are using may have some unwanted paths between nodes. To find your problem you must trace through the voltages in your circuit node by node and compare the signal you expect to have. Then if they are different use your engineering judgment to decide what is causing the different or ask your lab assistant.

# 1

# Open and Short Circuit Test

**Aim:** To perform open and short circuit test on a single-phase transformer, to obtain equivalent circuit parameters.

**Apparatus required:**

| S.No. | Device | Ratings | Type | Quantity |
|---|---|---|---|---|
| 1. | Autotransformer | (0-270) V/2.16 kVA | | 1 |
| 2. | Ammeter | (0-5) V | MI | 1 |
| 4. | Voltmeter | (0-250) V | MI | 1 |
| 5. | Ammeter | (0-10) A | MI | 1 |
| 6. | Wattmeter | LPF/UPF | ED type | 1 |
| 7. | Transformer | 0-240 V | 1-phase | 1 |

## Theory

### Open circuit test

The core loss or iron loss i.e. the no-load loss in a transformer is determined by performing the no-load test at normal sine wave impressed voltage. Through the iron loss in a transformer is not governed by the supply being connected to primary or secondary side but when measuring this no-load loss.

It is usually more convenient and safer to supply the voltage to the low voltage side. This makes the test apparatus cheaper and becomes safer for the low testing personal. To determine the shunt branch parameter of the equivalent circuit, the primary side (L.V.) is connected to a normal supply voltage and rated frequency and secondary side (H.V.) is left open with care so that there is no possibility of coming in contact with any part of (H.V.) terminals. The wattmeter connected in the primary (L.V.), will then read the total hysteresis.

The wattmeter connected in the primary (L.V.), will then read the total hysteresis and eddy current loss ($W_0$) in the single-phase transformer. Since the primary no-load current is small the copper losses due to it can be neglected.

$$VI_0 \cos \phi_0 = W_0 \qquad (1)$$

Where

$cos\ \phi_0$ = no-load power factor, $I_0$ = no-load current.

$$\phi_0 = \cos^{-1}\frac{W_0}{VI_0} \tag{2}$$

Also, from the no-load equivalent circuit shown in the figure

The shunt branch resistance

$$R_0 = \frac{V}{I_0 \cos\phi_0} \tag{3}$$

And

$$G_0 = \frac{I_0 \cos\phi_0}{V} \tag{4}$$

The shunt branch reactance

$$X_0 = \frac{V}{I_0 \sin\phi_0} \tag{5}$$

And

$$B_0 = \frac{I_0 \cos\phi_0}{V} \tag{6}$$

Where

$R_o$ = shunt branch resistance, $G_o$ = shunt branch conductance,

$X_o$ = shunt branch reactance, $B_o$ = shunt branch susceptance,

Thus, from no-load test, the shunt branch parameters (shunt branch conductance $G_0$ and shunt branch susceptance $B_o$) are known.

### Short circuit test

This test is performed on a single-phase transformer to find out the series parameters of the equivalent circuit as well as to obtain the full load copper loss. For this test short circuit of terminals are doe on the low-tension side of the transformer and adjust the voltage (by a factor of 5 to 20 lower than the rated voltage, depending on the type of transformer) on the other side so that full load current flows. It is preferable to short circuit low tension side (as the current rating of this side is higher). This loss during this short circuit will be full load copper losses in both the primary and secondary winding, the core loss being very small and is quite negligible. Precaution is to be taken about the magnitude of applied voltage at the H.T. side such that full load current not exceeded.

Therefore, if a wattmeter (U.P.F) is connection in the supply side it will read full load of the single-phase transformer. If,

$W_{sc}$ = Wattmeter reading,

$V_{sc}$ = Supply voltage at short circuit,

$I_{sc}$ = Full load current.

Then equivalent impedance referred to the primary side is,

$$Z_{sc} = \frac{V_{sc}}{I_{sc}} \tag{7}$$

Then equivalent resistance referred to the primary side is,

$$R_{sc} = \frac{W_{sc}}{I^2_{sc}} \tag{8}$$

Similarly,

$$X_{sc} = \sqrt{Z^2_{sc} - R^2_{sc}} \tag{9}$$

**Circuit diagram**

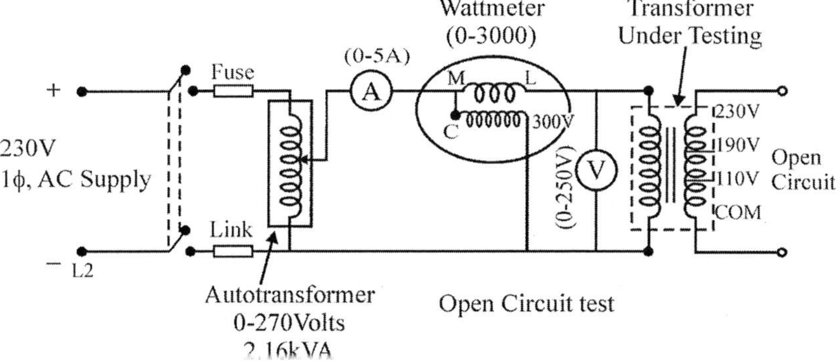

Open Circuit test

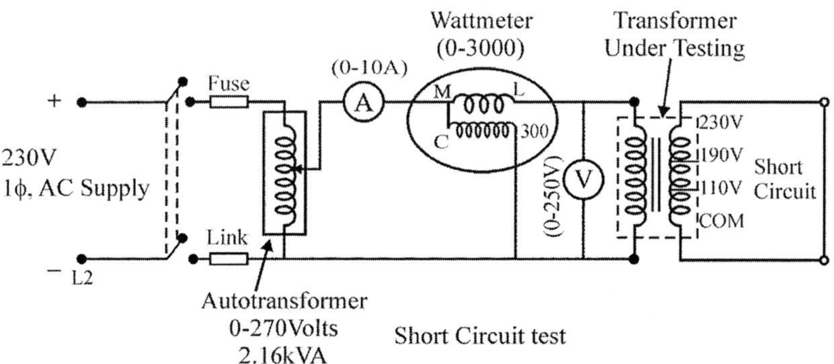

Short Circuit test

## Observation table

### Open circuit test

| S.No. | No-load current (**$I_0$**) | Applied voltage (**V**) | Wattmeter reading (**$W_0$**) |
|---|---|---|---|
| 1. | | | |
| 2. | | | |

### Short circuit test

| S.No. | Short circuit current (**$I_{sc}$**) | Applied voltage (**$V_{sc}$**) | Wattmeter reading (**$W_{sc}$**) |
|---|---|---|---|
| 1. | | | |
| 2. | | | |

## Calculation

### Open circuit test

$$X_0 = \frac{V}{I_0 \sin \phi_0} \qquad R_0 = \frac{V}{I_0 \cos \phi_0}$$

$$VI_0 \cos \phi_o = W_0 \qquad \phi_0 = \cos^{-1} \frac{W_0}{VI_0}$$

### Short circuit test

$$Z_{sc} = \frac{V_{sc}}{I_{sc}} \qquad R_{sc} = \frac{W_{sc}}{I^2_{sc}}$$

$$X_{sc} = \sqrt{Z^2_{sc} - R^2_{sc}}$$

## Result

The parameters of the equivalent circuit are found as below:

$R_o$ =

$X_o$ =

$X_{sco}$ =

$R_{sc}$ =

$Z_{sc}$ =

## Discussion

## Precautions

1. The connections should be neat, clean and tight.
2. Ammeter, Voltmeter should be of proper range.
3. Thick copper wires need to be used for connections, which have lower resistance that can be neglected.

## Viva-Questions

1) While performing a circuit test on transformer at 30v, 30 Hz it gives $\Phi_1$ as lagging power factor. If the same test is performed at 30V, 50Hz, then

   a) $\Phi 1 > \Phi 2$ b) $\Phi 1 < \Phi 2$

   c) $\Phi 1 = \Phi 2$ d) None of these

2) If the magnitude of leakage reactance is equal to the resistance of both primary and secondary of a single-phase transformer. Then the input power factor is

   a) $1/\sqrt{2}$ b) 1

   c) $\sqrt{2}$ d) Zero

3) Open circuit test is to be performed on transformer to find

   a) Constant losses

   b) Copper losses

   c) Both constant losses and copper losses

   d) None of these

4) If the transformer is step-up transformer, then for performing short circuit test meter is connected to the

   a) H.V. side

   b) L.V. side

   c) Any side

   d) Test cannot be performed on step up transformer

5) If short circuit test is performed on the transformer with constant rated voltage and increased frequency, then the short circuit current and power factor

   a) Both will increase b) Both will decrease

   c) Increase, decrease d) Decrease, remains constant

6) A single-phase transformer of 2200/220 V having rated l.v. current of 150 A has to undergo open circuit test on h.v. side. Which of the below instruments range should be used?

a) 6A, 200V

b) 150A, 22V

c) 60A, 220V

d) 6A, 20V

7) A single-phase transformer of 2200/220 V having rated l.v. current of 150 A has to undergo open circuit test on h.v side. The instruments used are voltmeter of 200V and ammeter of 1A. Then the results

a) Will be wrong

b) Will be accurate

c) Of ammeter will burn

d) None of the mentioned

8) Which of the following loss in a transformer is zero even at full load

a) Eddy current loss

b) Core loss

c) Copper loss

d) Friction loss

9) Which of the following test is performed to determine the leakage reactance

a) Short circuit test

b) Open circuit test

c) Both Open circuit and short circuit test

d) Test by an Impedance bridge

10) What would happen if a transformer is connected to a DC supply?

a) No effect

b) Operate with high efficiency

c) Damage the transformer

d) Operate with low frequency

# 2

# Direct Loading Method

**Aim:** Determination of **percentage regulation and efficiency** of a single-phase transformer by direct loading method.

## Apparatus required

| S.No. | Device | Ratings | Type | Quantity |
|---|---|---|---|---|
| 1. | Autotransformer | (0-270) V/2.16 kVA | | 1 |
| 2. | Loading lamp | (0-10) A | | 1 |
| 4. | Voltmeter | (0-250) V | MI | 2 |
| 5. | Ammeter | (0-10) A | MI | 1 |
| 6. | Wattmeter | | UPF | 2 |
| 7. | Transformer | | | 1 |

## Theory

### Percentage regulation

When the load on the transformer is increased, the voltage drop in its primary and secondary windings also increases causing variations in secondary terminal voltage. The term voltage regulation of transformer means the change in secondary terminal voltage when full load is thrown off keeping supply voltage and frequency constant. It is generally expressed as percentage of no-load terminal voltage for a desired power factor.

$$\text{Voltage regulation}=\frac{\text{No load voltage-Full load voltage}}{\text{Full load voltage}}$$

$$\%\text{Regulation}=\frac{\left(E_2\text{-}V_2\right)}{V_2}*100$$

Where

$E_2$ = No-load voltage (volts), $V_2$=Full load voltage (volts)

The voltage regulation is an important measure of transformer performance. The limit of voltage variation is specified in term of voltage regulation. For example, transformer in public supply system must be so adjusted that the voltage at the terminals of the consumers must not exceed ± 5%.

### Transformer efficiency (η)

The efficiency of a transformer at a particular load and power factor is defined as the output divided by input (the two being measured in the same unit).

$$\text{Transformer efficiency } (\eta) = \frac{\text{Output power}}{\text{Input power}}$$

Since, the transformer is a static device there are no rotational losses such as wind age and frictional losses in a rotating machine. In a well-designed transformer the efficiency can be as high as 99%.

One simple method is to connect wattmeter in primary and secondary circuits and measure input and output power and then calculate the efficiency.

## Circuit diagram

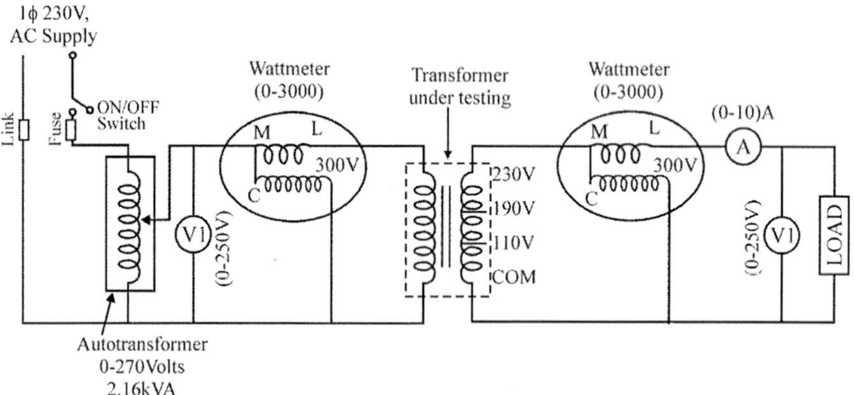


## Observation table

No-load voltage ($E_2$) = ………… Volts

| S.No. | Load | Input Voltage ($V_1$) | Input Power ($W_1$) | Output Power ($W_2$) | Load Current ($I_2$) | Load Voltage ($V_2$) |
|---|---|---|---|---|---|---|
| | | | | | | |
| | | | | | | |
| | | | | | | |
| | | | | | | |
| | | | | | | |
| | | | | | | |
| | | | | | | |

## Calculation

$$\% \text{ Efficiency} = \frac{\text{Output power}}{\text{Inputpower}} * 100$$

$$\eta = \frac{W_2}{W_1} * 100$$

$$\%\text{Regulation} = \frac{(E_2 - V_2)}{V_2} * 100$$

## Result

The average % efficiency of the transformer is %

Percentage Regulation is different for different values of load connected.

## Model Graphs

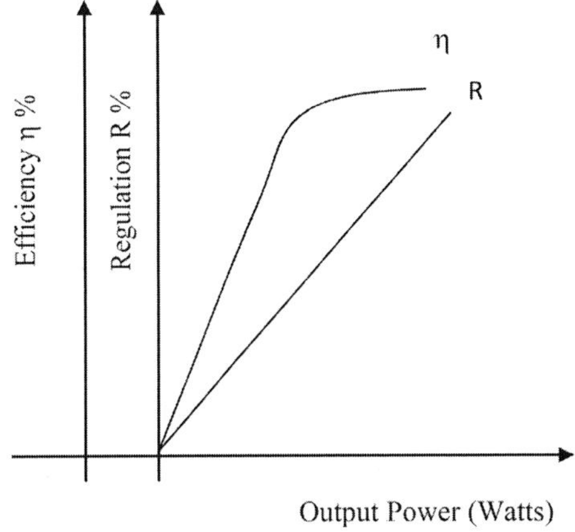


## Discussion

## Precautions

1. Auto Transformer should be in minimum position.
2. The AC supply is given and removed from the transformer under no-load condition

## Viva-Questions

1) A transformer can have zero voltage regulation at

a) Leading power factor
b) Lagging power factor
c) Unity power factor
d) Zero power factor

2) Negative voltage regulation indicates

a) Capacitive loading only
b) Inductive loading only
c) Inductive or resistive loading
d) Cannot be determined

3) A good voltage regulation of a transformer indicates

a) Output voltage fluctuation from no-load to full load is least
b) Output voltage fluctuation with power factor is least
c) Difference between primary and secondary voltage is least
d) Difference between primary and secondary voltage is maximum

4) The efficiency of a transformer will be maximum when

a) Copper losses = hysteresis losses
b) Hysteresis losses = eddy current losses
c) Eddy current losses = copper losses
d) Copper losses = iron losses

5) In a given transformer for given applied voltage, losses which remain constant irrespective of load changes are

a) Friction and windage losses
b) Copper losses
c) Hysteresis and eddy current losses
d) None of the above

6) In a transformer routine efficiency depends upon

a) Supply frequency
b) Load current
c) Power factor of load
d) Both (b) and (c)

7) A 10 kVA, 400/200 V, 1-phase transformer with 2% resistance and 2% leakage reactance. It draws steady short circuit current at angle of

a) 45°
b) 75°
c) 135°
d) 0°

8) While conducting testing on the single-phase transformer, one of the student tries to measure the resistance by putting an ammeter across one terminal of primary and other to secondary, the reading obtained will be

a) Infinite
b) Zero
c) Finite
d) negative finite

9) A 600 kVA transformer has iron losses of 400 kW and copper losses of 500 kW. Its kVA rating for maximum efficiency is given by

a) 537 kVA
b) 548 kVA
c) 555 kVA
d) 585 kVA

10) If the power factor of a transformer is increases, then its efficiency will

a) Increases
b) Decreases
c) Remains constant
d) Not related to each other

# 3

# Sumpner's Test

**Aim:** To perform Sumpner's test (back-to-back test) on a pair of similar transformer and thus calculating efficiency of operation.

## Apparatus required

| S.No. | Device | Rating / Range | Type | Quantity |
|---|---|---|---|---|
| 1. | Autotransformer | (0-270) V /2.16 kVA | --- | 2 |
| 2. | Voltmeter | (0-100) V | MI | 1 |
| 4. | Voltmeter | (0-250) V | MI | 1 |
| 5. | Ammeter | (0-2) A. | MI | 1 |
| 6. | Ammeter | (0-5) A. | MI | 1 |
| 7. | Wattmeter | LPF /UPF | ED type | (1/1) |
| 8. | Transformer | (1 :1) kVA | --- | 2 |

## Theory

Sumpner's test givens an effective way, to actually see the performance of transformers of large rating without loading them, at full load. This back to back test allows heat run test so that temperature rise and its rate can be analysed.

The two identical transformers are connected back to back as shown in figure such that their polarities are in opposition. To do this, two terminals of the secondary winding are connected as shown and the voltages between the other two are measured by exciting primaries.

The reading would be either zero or twice of the secondary voltage. If the reading is zero the back to back connection is correct otherwise reverse the connection of one of the secondary.

With the secondary so connected if the two free terminals are connected, the ammeter will show zero current i.e., no circulating current only the magnetizing current will be drawn from the supply, which will be responsible for the core losses of the two transformers.

Low voltage regulated supply is needed for circulating a current around the primary circuit. Thus by regulating this low voltage booster supply full load current can be established in the primary.

The wattmeter now connected in the secondary will read the total copper losses in the winding of the two transformers.

## Procedure

1. Make connections as shown in the figure.
2. Connected two terminals first (say b' and c') and check the voltage across a' and d' by a double range voltmeter.
3. Both the primaries are connected in parallel and with normal voltage (rated).impressed across it.
4. Adjust the secondary induced voltage so the circulating current is equal to the secondary winding current.
5. Write down the values of current, voltage and power of both the secondary and primary winding.

## Circuit diagram

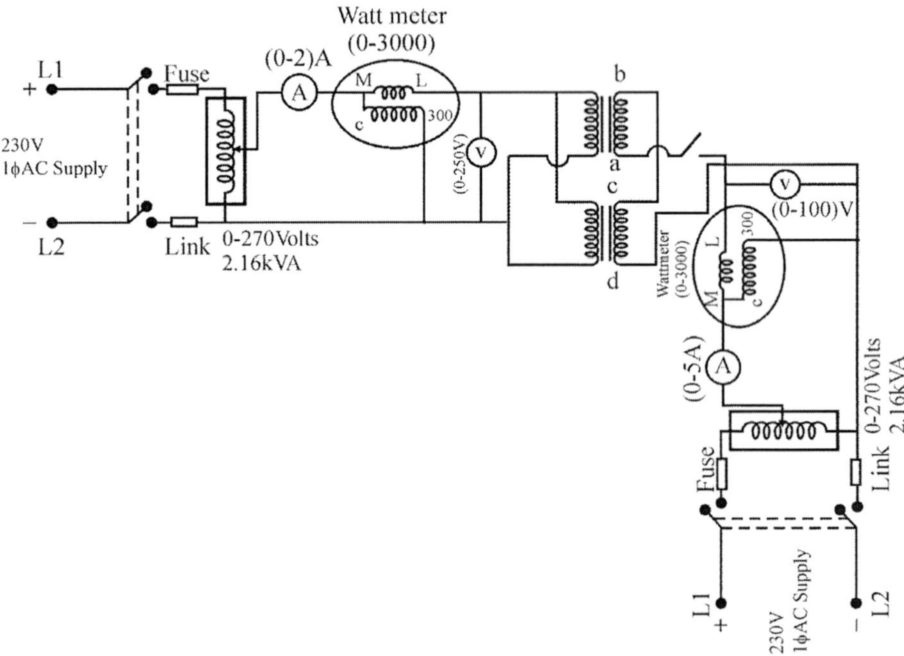

## Observation table

| S.No. | Current in primary winding ($I_1$) | Voltage in primary winding ($V_1$) | Power in primary winding ($W_1$) | Current in secondary winding ($I_2$) | Voltage in secondary winding ($V_2$) | Power in secondary winding ($W_2$) |
|---|---|---|---|---|---|---|
| 1. | | | | | | |
| 2. | | | | | | |

## Note

$W_2$ will give the copper losses in the winding of both the transformers $(W_1+W_2)/2$ will give the total core and copper losses in one transformer calculate the efficiency of the transformer from the above readings.

## Calculation

Core loss of each transformer $W_o = \frac{W_1}{2}$ Watts (1)

Full load copper loss of each transformer $W_{sc} = \frac{W_2}{2}$ Watts (2)

$$W_o = V_1 I_1 Cos\phi_o \tag{3}$$

$$\phi_o = Cos^{-1}\frac{W_o}{V_1 I_1} \tag{4}$$

$$I_1 = \frac{I_o}{2} \tag{5}$$

$$I_w = I_1 Cos\phi o \tag{6}$$

$$I_\mu = I_1 Cos\phi o \tag{7}$$

$$R_o = \frac{V_1}{Iw} \tag{8}$$

$$X_o = \frac{V_1}{I\mu} \tag{9}$$

$$R_{o2} = \frac{Wc}{I_2^2} \tag{10}$$

$$Z_{o2} = \frac{V_2}{I_2} \tag{11}$$

$$X_{o2} = \sqrt{Z_{o2}^{\ 2} - R_{o2}^{\ 2}} \tag{12}$$

Copper loss at various loads = $I_2^{\,2}R_{o2}$

$$\text{Efficiency} = \frac{\text{Output power}}{\text{Output power} + \text{Losses}}$$

## Result

Thus, the efficiency and regulation of a given single-phase Transformer is carried out by conducting back-to-back test and the equivalent circuit parameters are found out.

## Discussion

## Precautions

1. Auto Transformer who's variac should be in zero position, before switching on the ac supply.
2. Transformer should be operated under rated values.

## Viva-Questions

1) Sumpner's test is conducted on transformers to study effect of

   a) Temperature  b) Stray losses

   c) All-day efficiency  d) Cannot be determined

2) Which of the following tests are enough to find all the parameters related to a transformer?

   a) OC test  b) OC, SC test

   c) OC, SC, Sumpner's test  d) Sumpner's test

3) Sumpner's test is performed on

   a) Single transformer at a time  b) Only two transformers at a time

   c) Minimum 2 transformers at a time  d) Many transformers at a time

4) In Sumpner's test

   a) Primaries can be connected in either way

   b) Primaries are connected in parallel with each other

   c) Primaries of both transformers are connected in series with each other

   d) No need to connect primaries

5) Which test is sufficient for efficiency of two identical transformers under load conditions?

a) Short-circuit test
b) Back-to-back test
c) Open circuit test
d) Any of the above

6) In Sumpner's test

a) Two secondaries are connected in phase opposition
b) Two secondaries are connected in phase addition
c) Can be connected in either way
d) Never connected with each other

7) When secondaries are connected in phase opposition, power drawn by the circuit is equal to

a) $2P_i$
b) $P^2_i$
c) $P_i$
d) $2P_C$

8) When the AC supply at the primary side of a transformer are shorted, power drawn by the circuit is equal to

a) $2P_C$
b) $2P_i$
c) $2P_C + 2P_i$
d) can't be determined

9) Total power required for Sumpner's test is given by

a) $P_C + P_i$
b) $P_C + 2P_i$
c) $2P_C + P_i$
d) $2(P_C + P_i)$

10) When the AC supply at the primary side of a transformer are shorted, power drawn by the circuit is equal to

a) $2P_C$
b) $2P_i$
c) $2P_C + 2P_i$
d) Can't be determined

# 4

# Parallel Operation of Transformer

**Aim:** Perform the parallel operation of single-phase transformer and observe the load sharing of each transformer.

## Apparatus required

| S.No. | Apparatus | Type | Range | Quantity |
|---|---|---|---|---|
| 1 | Voltmeter | MI | 0-300 V | 1 |
| 2 | Ammeter | MI | 0-10 A | 2 |
| 3 | Ammeter | MI | 0-20 A | 1 |
| 4 | Single-phase Variac | Variable | 230/0-270 | 1 |
| 5 | Single-phase Transformer | Step-up | 3 kVA, (115/230) | 2 |
| 6 | Resistive Load | | 5 kW, 220 V | 1 |

## Theory

A power transformer is one of the most vital and an equally expensive component in a power system. It may so happen that, over time, due to load growth in its service area, an existing transformer may not be able to withstand the demand during peak-hours without exceeding its long-term MVA rating. Operating a transformer in such a fashion would cause overheating and degrade its expected life. In most cases, instead of commissioning an entirely new higher capacity unit, a more viable alternative exists in adding a smaller unit in parallel to complement the existing one. In other words, a new smaller capacity transformer can now be connected in parallel to the existing one such that the two shares a large peak load in a specific proportion and the one operating near limits is relieved of the burden.

Also, during light load conditions, the additive capacity can be kept offline, if desired. To successfully operate the transformers in parallel, while commissioning, certain rules must be followed. These are as follows-

1. Both the transformers must have same polarities.
2. Both the transformers must have same voltage ratings.
3. Both the transformers must have the same impedance percentage.
4. The circulating current must be zero.

Now,

The total load current is given as

$$I = I_1 + I_2$$

$$I_1Z_1 = 1_2Z_2$$

$$I_1 = I\frac{Z_2}{Z_1 + Z_2}$$

$$I_2 = I\frac{Z_1}{Z_1 + Z_2}$$

$$S_1 = V^* I_1$$

$$S_2 = V^* I_2$$

$$\frac{S_1}{S_2} = \frac{I_1}{I_2} = 1 = \frac{Z_1}{Z_2}$$

Using the above expression, the load sharing can be found.

## Circuit Diagram

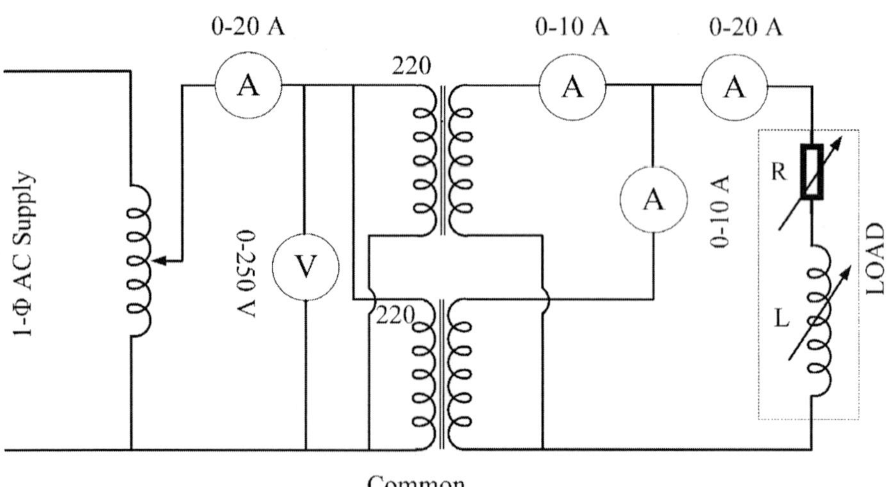


**Parallel Operation of Transformers**

## Observation Table

| S.No. | Voltage | I(load) | $I_1$ | $I_2$ | $I_3 = I_1 + I_2$ |
|---|---|---|---|---|---|
| | | | | | |
| | | | | | |
| | | | | | |

## Calculation

$S_1 = V^* I_1$

$S_2 = V^* I_2$

As per the expression, $S_1$ and $S_2$ can be calculated at different load levels.

## Conclusion

The parallel operation on the two transformers is performed and the amount of load shared by each is obtained.

## Discussion

## Viva-Questions

1. Two transformers are connected in parallel, not having unequal percentage impedances, which statement is correct?
   a) Short-circuiting of the secondary's
   b) Power factor of one of the transformers is leading while that of the other lagging
   c) Transformers having higher copper losses will have negligible core losses
   d) Loading of the transformers not in proportion to their kVA ratings
2. For the parallel operation of two single-phase transformers it is necessary that they should have
   a) same efficiency
   b) Same polarity
   c) Same kVA rating
   d) Same number of turns on the secondary side
3. Transformers operating in parallel mode of operation will share the load depending upon their
   a) Leakage reactance
   b) Per unit impedance
   c) Efficiencies
   d) Ratings

5. What will happen if the transformers working in parallel are not connected with regard to polarity

   a) The power factor of the two trans-formers will be different from the power factor of common load

   b) Incorrect polarity will result in dead short circuit

   c) The transformers will not share load in proportion to their kVA ratings

   d) Cannot be determined

5. If the percentage impedances of the two transformers working in parallel are different, then

   a) Transformers will be overheated

   b) Power factors of both the transformers will be same

   c) Parallel operation will be not possible

   d) Parallel operation will still be possible, but the power factors at which the two transformers operate will be different from the power factor of the common load.

6. While connecting two transformers in parallel voltage around the local loop

   a) Positive b) Negative

   c) Equals zero d) Insufficient information

7. Y/Y and Y/D transformers can be paralleled.

   a) True b) False

8. Define circulating current?

9. What is the effect of circulating current when transformers are connected in parallel?

10. What do you mean by impedance of a transformer?

# 5

# Different Connections of 3-Phase Transformer

**Aim:** To verify the relationship between the input and output voltage for 3-phase transformer connection in various ways:

a) Delta-Star
b) Delta-Delta
c) Star-Delta
d) Star-Star

## Apparatus required

| S.No. | Device | Type | Range | Quantity |
|---|---|---|---|---|
| 1 | Voltmeter | MI | 0-600 V | 1 |
| 2 | Voltmeter | MI | 0-300 V | 1 |
| 3 | Transformer | - | 3- phase | 1 |

## Theory

A 3-phase transformer can be considered as a set of 3 single-phase transformer bank or as single unit of 3 phase windings. Since three winding are available in both primary and secondary side then there can be different types of connections and so far, different characteristics.

### 1) Delta-Star Connection ($\Delta - Y$)

In $\Delta - Y$ connection, the primary line voltage is equals to the phase voltage. In secondary side the line voltage is $\sqrt{3}$ times the phase voltage as secondary is connected in star configuration. Therefore, the ration of line-to-line voltage in primary and secondary is given by

$V_{LP} = V_{PP}$ $\qquad$ $V_{LS} = \sqrt{3}V_{PS}$

Then $\dfrac{V_{LP}}{V_{LS}} = \dfrac{a}{\sqrt{3}}$

Where,

$$a = \frac{V_{PP}}{V_{PS}}$$

**2) Delta-Delta connection ($\Delta - \Delta$)**

In $\Delta - \Delta$ connection, the primary line voltage is equals to the primary phase voltage $V_{LP} = V_{PP}$. The relation between secondary line and phase voltage is same as in primary as both are $\Delta$ connected. Therefore, the ration of line-to-line voltage in primary and secondary is given by

$$V_{LP} = V_{PP}$$

$$V_{LS} = V_{PS}$$

Then $\frac{V_{LP}}{V_{LS}} = a$

Where $a = \frac{V_{PP}}{V_{PS}}$

**3) Star-Delta ($Y - \Delta$)**

In $Y - \Delta$ connection, the primary line voltage is equal to the $\sqrt{\ }$ times the primary phase voltage $V_{LP} = \sqrt{3}V_{PP}$. The secondary line voltage is equal to the secondary phase voltage as it is in delta configuration. Therefore, the ration of line-to-line voltage in primary and secondary is given by

$$V_{LP} = \sqrt{3}V_{PP}$$

$$V_{LS} = V_{PS}$$

$$\frac{V_{LP}}{V_{LS}} = \sqrt{3}a$$

Where $a = \frac{V_{PP}}{V_{PS}}$

**4) Star-Star connection ($Y - Y$)**

In $Y - Y$ connection, the primary line voltage is equal to the $\sqrt{3}$ times the primary phase voltage $V_{LP} = \sqrt{3}V_{PP}$. In secondary side the line voltage is $\sqrt{3}$ times the phase voltage as secondary is connected in star configuration. Therefore, the ration of line-to-line voltage in primary and secondary is given by

$$V_{LP} = \sqrt{3}V_{PP}$$

$$V_{LS} = \sqrt{3}V_{PS}$$

$$\frac{V_{LP}}{V_{LS}} = a$$

Where $a = \frac{V_{PP}}{V_{PS}}$

## Observation Table

The primary to secondary turn ratio (a) = 3:1

### 1) Delta-Star Connection (Δ – Y)

| S.No. | Line voltage at primary side ($V_{LP}$) | Line voltage at secondary side ($V_{LS}$) | Theoretical value of Line voltage at secondary side ($V_{LS}$) |
|---|---|---|---|
| | | | |

### 2) Delta-Delta connection (Δ –Δ)

| S.No. | Line voltage at primary side ($V_{LP}$) | Line voltage at secondary side ($V_{LS}$) | Theoretical value of Line voltage at secondary side ($V_{LS}$) |
|---|---|---|---|
| | | | |

### 3) Star-Delta (Y – Δ):

| S.No. | Line voltage at primary side ($V_{LP}$) | Line voltage at secondary side ($V_{LS}$) | Theoretical value of Line voltage at secondary side ($V_{LS}$) |
|---|---|---|---|
| | | | |

### 4) Star-Star connection (Y - Y):

| S.No. | Line voltage at primary side ($V_{LP}$) | Line voltage at secondary side ($V_{LS}$) | Theoretical value of Line voltage at secondary side ($V_{LS}$) |
|---|---|---|---|
| | | | |

## Result

The line voltage at secondary side is calculated and also found by experiment and both are approximately equal.

## Discussion

## Circuit Diagram

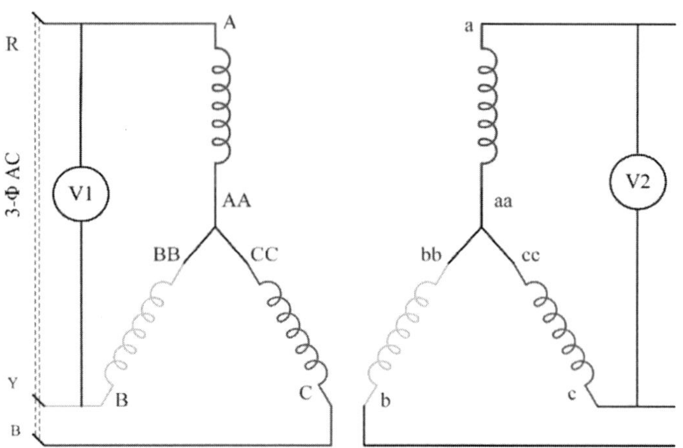


**Fig. 1:** Star-Star Connection of 3-phase Transformer

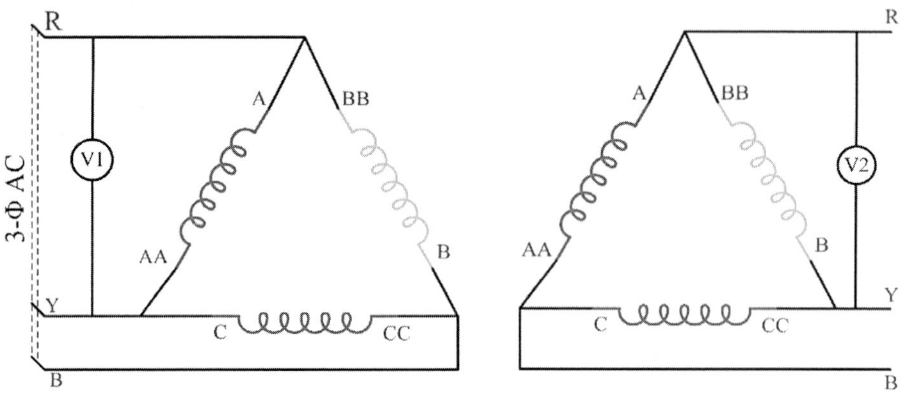


**Fig. 2:** Delta-Delta Connection of 3-phase Transformer

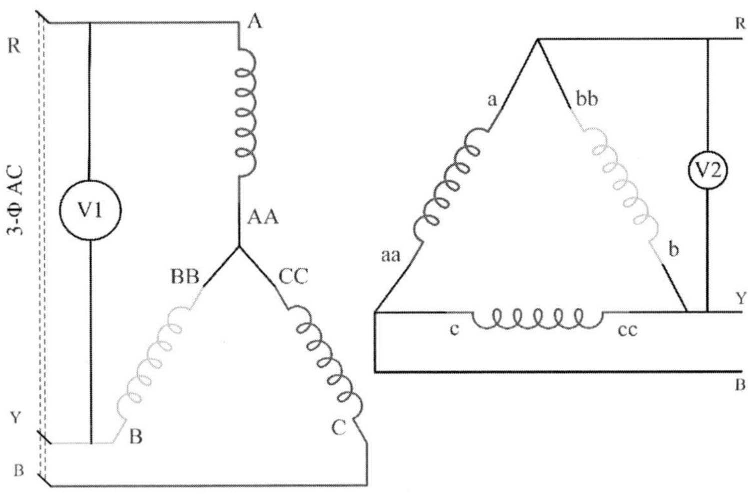


**Fig. 3:** Star-Delta Connection of 3-phase Transformer

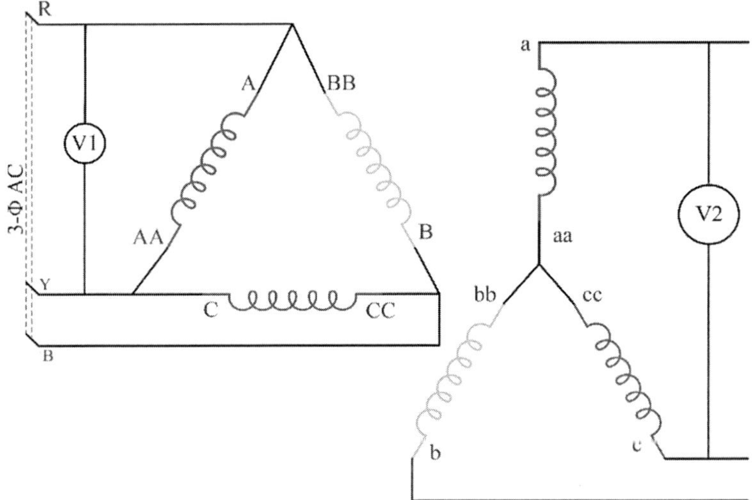


**Fig. 4:** Delta-Star Connection of 3-phase Transformer

## Viva-Questions

1. A V-V connected transformer can be connected in parallel to delta-delta connected transformer but not to

   a) Delta-star  b) Star-delta

   c) Star-V  d) All of the mentioned

2. Three units of single-phase transformers and one single three-phase transformer rating
   - a) Will be same for one rating
   - b) Can never be made same
   - c) May be same
   - d) None of the mentioned
3. The transformer which is more feasible to use in the distribution ends should be
   - a) Star-delta
   - b) Delta-star
   - c) Scott
   - d) Delta-delta
4. The delta-delta connections are used in applications of
   - a) Large l.v. transformers
   - b) Small h.v. transformers
   - c) Large h.v. transformers
   - d) Small l.v. transformers
5. Open delta transformers can be obtained from
   - a) Delta-delta
   - b) Star-delta
   - c) Delta-star
   - d) Any of the mentioned
6. If one of the transformers is removed from the bank of only delta-delta, then it behaves with 58% power delivery.
   - a) True
   - b) False
7. Shape of emf generated by delta connected transformer is not always sinusoidal.
   - a) True
   - b) False
8. The secondary line to line voltage of a star-delta connected transformer is measured to be 400 V. If the turns ratio between the primary and secondary coils is 2:1, the applied line to line voltage in the primary is:
   - (a) 462 V
   - (B) 346 V
   - (b) 1386 V
   - (D) 800 V

# 6

# Scott Connection of Transformers

**Aim:** To obtain balanced two-phase supply from three-phase supply by Scott arrangement of two single-phase transformers.

## Apparatus Required

| S.No. | Name | Type | Range | Quantity |
|---|---|---|---|---|
| 1 | Voltmeter | MI | 0-600 V | 1 |
| 2 | Voltmeter | MI | 0-300 V | 1 |
| 3 | Ammeter | MI | 0-5 A | 5 |
| 4 | 3 phase Variac | - | 420/0-450 V | 1 |
| 5 | Lamp bank load | Resistive | 250 v, 5 kW | 1 (dual output) |

## Theory

The phase conversion from three to two phase is needed in special cases, such as in supplying 2-phase electric arc furnaces. Scott connection of two-single-phase transformers is employed for conversion of a three-phase system to two phase system or vice-versa. Rating of one transformer should be 15% greater than that of the other, but in practical two identical transformers are used for interchangeability and spares. The 50% tap of one transformer (Main transformer) is connected to 86.6% tap of the other transformer (Teaser transformer). The secondaries for balanced supply system have equal number of turns.

Consider the Scott connection of two single-phase transformers with turn's ratio N1:N2 as shown below. The phase diagram of line voltages on the primary side, $V_{AB}$, $V_{BC}$, $V_{CA}$ form an equilateral triangle.

$$V_{AB} = V\angle 0^0$$

$$V_{BC} = V\angle -120^0$$

$$V_{CA} = V\angle 120^0$$

The secondary voltage of the main transformer is given by,

$$V_b = \frac{N_2}{N_1}V_{CA} = \frac{N_2}{N_1}V\angle 60^0$$

The voltage $V_{AM}$ is given by,

$$V_{AM} = V_{AB} + \frac{V_{BC}}{2}$$

$$V_{AM} = V + \frac{V\angle -120^0}{2}$$

$$V_{AM} = V\left(\frac{3}{4} - j\frac{\sqrt{3}}{4}\right)$$

$$V_{AM} = \frac{\sqrt{3}}{2} V\angle -30^0$$

This voltage VAM is across (√3/2) *N1 turns. Therefore, the primary voltage (across N1 turns) of teaser transformer is given by,

$$V_{AA'} = \frac{2}{\sqrt{3}} V_{AM}$$

$$V_{AA'} = V\angle -30^0$$

Hence, the secondary voltage of the teaser transformer is given by,

$$V_a = \frac{N_2}{N_1} V_{AA'} = \frac{N_2}{N_1} V\angle -30^0$$

Hence from the expression of $V_a$ and $V_b$, it can be observed that form a balanced 3 phase supply at input, balanced 2 phases can be obtained with a phase displacement of $90^0$.

If the secondary load currents are $I_a$ and $I_b$, then the primary currents can be obtained as follows,

$$I_A = \frac{2N_2}{\sqrt{3}N_1} I_a$$

$$I_{CB} = \frac{N_2}{N_1} I_b$$

$$I_B = -I_{CB} - \frac{I_A}{2}$$

$$I_C = I_{CB} - \frac{I_A}{2}$$

Note that for balanced load, the two secondary currents are equal (Ia=Ib) in magnitude and the three primary currents are also equal (IA=IB=IC) in magnitude. To get single-phase voltage supply, short negative polarity side of teaser transformer and positive polarity of main transformer on secondary side and take the voltage across positive polarity of teaser transformer and negative polarity of main transformer on secondary side.

This single-phase voltage is given by,

$$V = \sqrt{V_a^{\,2} + V_b^{\,2}}$$

Since, $V_a$ and $V_b$ are 90 degrees apart. In case of single-phase configuration, the secondary currents of teaser and main transformer are same i.e.

$$I_a = I_b$$

## Circuit Diagram

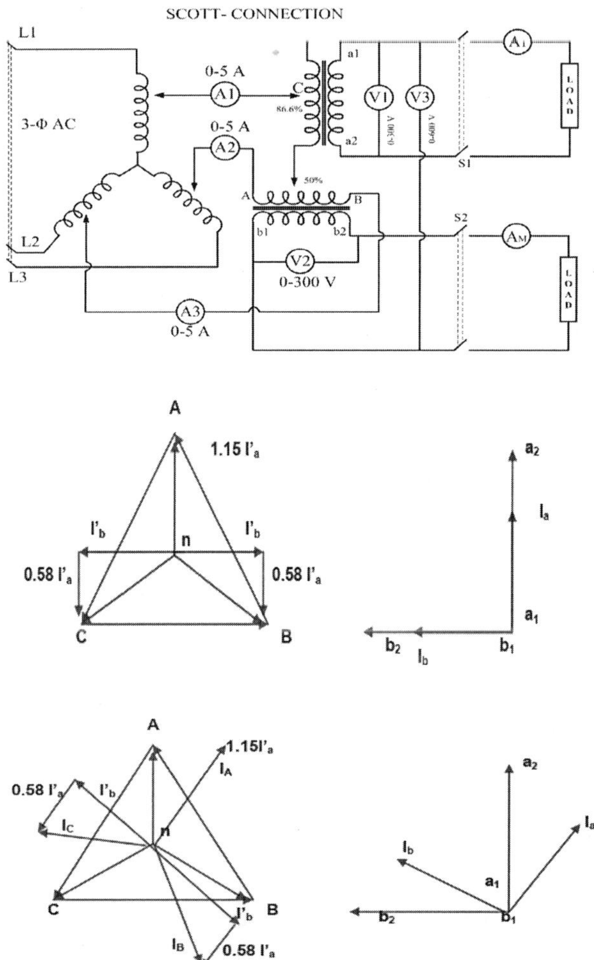


**Fig. 2:** Phasor diagrams **(a)** Balanced unity power factor load and **(b)** Unbalanced load

## Observation Table

| for balanced two phase supply | | | | under load conditions | | | | | | | |
|---|---|---|---|---|---|---|---|---|---|---|---|
| S.No. | $V_1$ | $V_2$ | $V_3$ | S.No. | $I_{2M}$ | $I_{2T}$ | $I_1$ | $I_2$ | $I_3$ | $V_1$ | $V_2$ |
| | | | | | | | | | | | |

## Result:

$I_A = I_B = I_C =$...................... Amp

$V_{AB} = V_{BC} = V_{CA} =$ ........... Volts

## Discussion

## Precautions

1. All connections should be tight and correct.
2. Switch off the supply when not in use.

## Viva-Questions

1. Only Scott connection is used for
   a) Converting three-phase to two-phase conversion
   b) Converting three-phase to single-phase conversion
   c) Converting single-phase to two-phase conversion
   d) None of the mentioned
2. In Scott connection, according to the vector diagram two windings are placed at
   a) At 1200 to each other
   b) Perpendicular with respect to each other
   c) At 600 to each other
   d) Can't say
3. What is the ratio of voltage/turn of two windings of a transformer in operated in Scott connection?
   a) ½    b) 1/√2
   c) 1    d) 1/√3

4. The primaries of two transformers in Scott connection are in turns ratio of

a) $\sqrt{3}/2$: 1
b) $2/\sqrt{3}$: $\sqrt{2}$
c) 1:1
d) Can't say

5. The secondaries of two transformers in Scott connection are in ratio of

a) 1:1
b) $\sqrt{3}/2$: 1
c) $\sqrt{3}$: 1
d) $\sqrt{2}$: 1

6. In Scott connection neutral point is located at

a) Three phase side at teaser
b) Two-phase side
c) Three phase side secondary
d) Anywhere

7. For single-phase to three-phase change in transformer which of the following connection is most suitable?

a) Scott connection
b) Scott connection with resistor network
c) Scott connection with some energy storing device
d) Any of the mentioned

8. For 3/6-phase converter transformers

a) Scott connection is used
b) Each secondary winding is divided
c) Each primary winding is divided
d) Can't be done

# 7

# Load Characteristics

**Aim:** To determine the Load characteristics (External characteristics) of self-excited D.C. shunt generator.

## Apparatus required

| S.No. | Device | Rating / Range | Type | Quantity |
|---|---|---|---|---|
| 1. | Shunt generator | | --- | 1 |
| 2. | Ammeter | (0-5) A | MC | 2 |
| 3. | Loading Rheostat / lamp | (0-10)A | --- | 1 |
| 4. | Voltmeter | (0-250) V | MC | 2 |
| 5. | Tachometer | (0-2000) RPM | Digital | 1 |
| 6. | Rheostats | (145 Ω/2.8A) | --- | 2 |

## Theory

If a dc shunt generator, after building up to its voltage, were loaded, its terminal voltage will drop. This drop increases if load increases. But this type of terminal drop is undesirable for a specified service. The relation between terminal voltage and load current is called external characteristics. Speed is to be kept constant. Reason for voltage drop:

1. Due to armature reaction.
2. Due to armature resistance drop.
3. Decrease of field current due to reduction in armature voltage.

The aim of this test also is to determine $\eta$, regulation and temperature rise at different loads.

## Circuit diagram

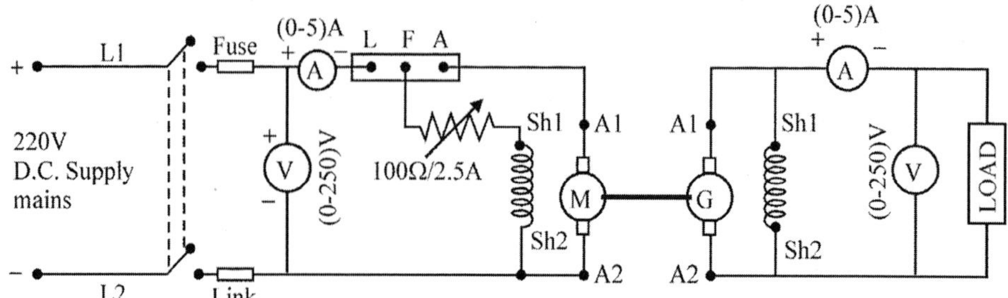


Load Characteristics of DC Shunt Generator

## Procedure

1. Make connections as shown in the figure.
2. Start dc motor by starter and adjust by field rheostat to rated speed.
3. Adjust the field resistance of dc shunt generator at a value to obtain 220 volts dc at no-load at rated speed of the dc generator.
4. Measure current (I) through the field regulator and voltage across it note them in the table given.
5. Connect a loading rheostat (lamp) as shown through a 2-pole switch fuse.
6. Repeat the above procedure and fill up the table.

## Observation table

Generator field resistance = (to be kept constant)

No-load speed=

No-load voltage=

| S.No. | Supply Voltage (volts) | Supply Current (Amp.) | Generator Load Current ($I_l$) (Amp.) | Terminal (generator) voltage ($V_t$) (volts) | Generator Speed (RPM) |
|---|---|---|---|---|---|
| 1. | | | | | |
| 2. | | | | | |
| 3. | | | | | |
| 4. | | | | | |
| 5. | | | | | |

**Calculation:** As per theory given above.

## Result

Draw the load (External) characteristic [Generator terminal voltage (**$V_L$**) Vs Load current (**$I_L$**)].

## Discussion

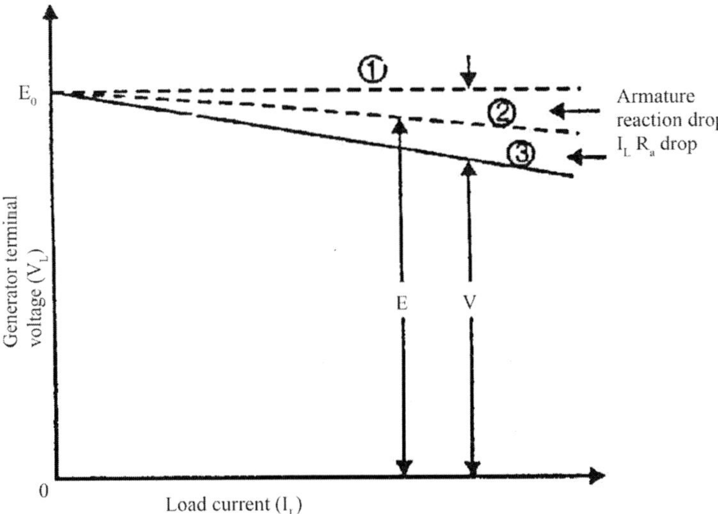


## Precautions

1. The field rheostat of motor should be at minimum position.
2. The field rheostat of generator should be at maximum position.
3. No-load should be connected to generator at the time of starting and stopping.

## Viva-Questions

1) Which of the following characteristics reveal about the magnetization nature of the machine?

   a) No-load characteristics

   b) Load characteristics

   c) Armature characteristics

   d) Both no-load and load characteristics

2) Choose the most inappropriate out of the following for the no-load characteristics of the dc generator.

   a) It is the open circuit characteristic of the machine
   b) It is magnetization characteristic of the machine
   c) It is conducted on the unloaded machine
   d) None of the mentioned

3) The external characteristic is plotted between

   a) terminal voltage vs armature current at constant excitation
   b) terminal voltage vs field current at constant armature current
   c) induced armature emf vs armature current at constant excitation
   d) none of the mentioned

4) A student forgot to mark the x-y axes in his experiments but he just noted down the cause and the effect for each. How will he conclude about the armature characteristic out of the all plotted graphs?

   a) By marking graph for constant terminal voltage
   b) By marking graph for constant field current
   c) By marking graph for constant armature current
   d) By marking graph for constant speed

5) Armature characteristic is also known as

   a) regulation characteristic
   b) magnetization characteristic
   c) external characteristic
   d) load characteristic

6) The air gap line represents

   a) Magnetic behavior of the air gap of the dc machine
   b) Magnetic behavior of the air gap of the induction machine
   c) Magnetic behavior of the iron core
   d) all of the mentioned

7) Identify the armature characteristic of the dc generator.

a)

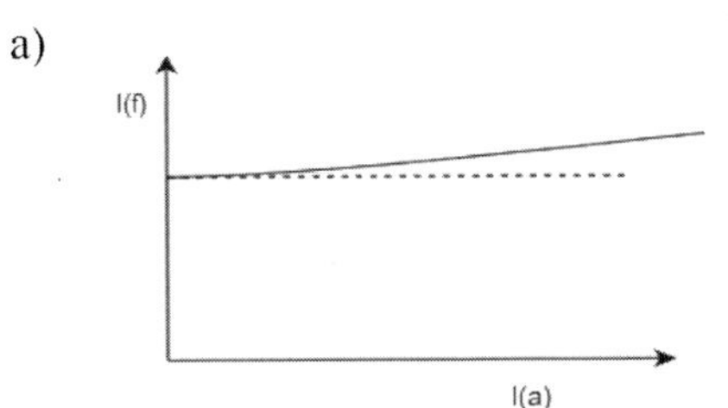


b)

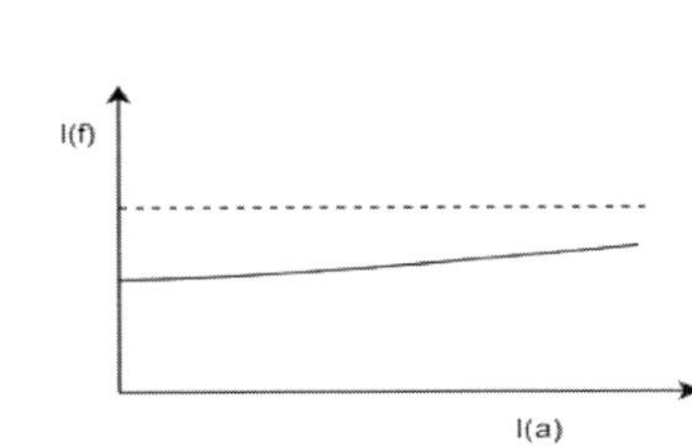


c)

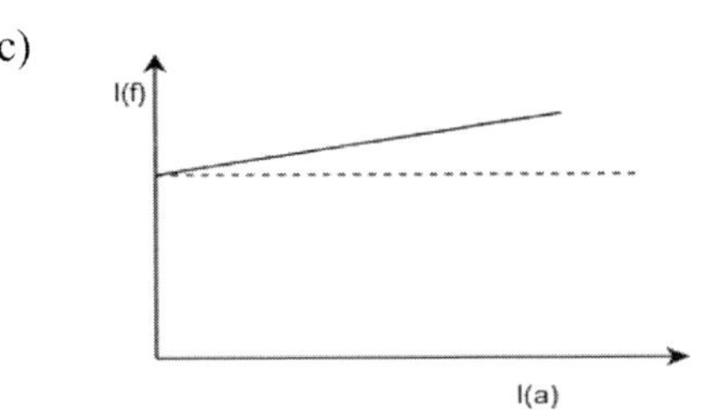


d)

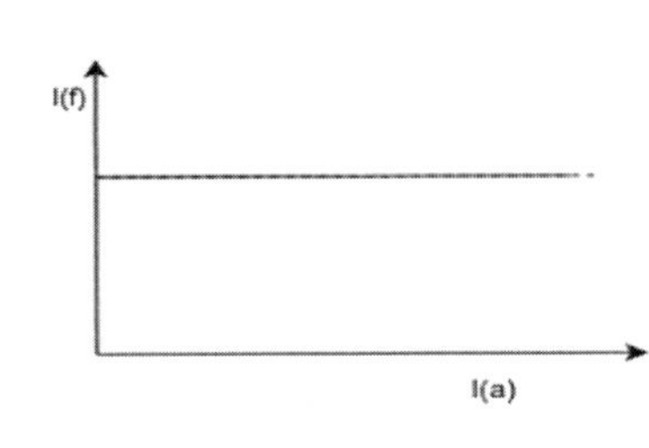


8) For a given dc generator, the external characteristic is plotted . Without using further plots, how can we obtain internal characteristic?

a) By adding the IaRa drop to the plot b) By adding armature reaction

c) By reducing IaRa drop d) All of the mentioned

9) Identify the machines by observing their external characteristics for (i) and (ii) respectively.

c)

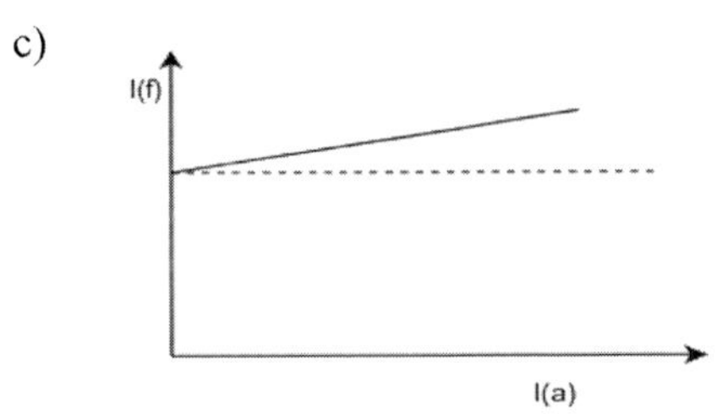


d)

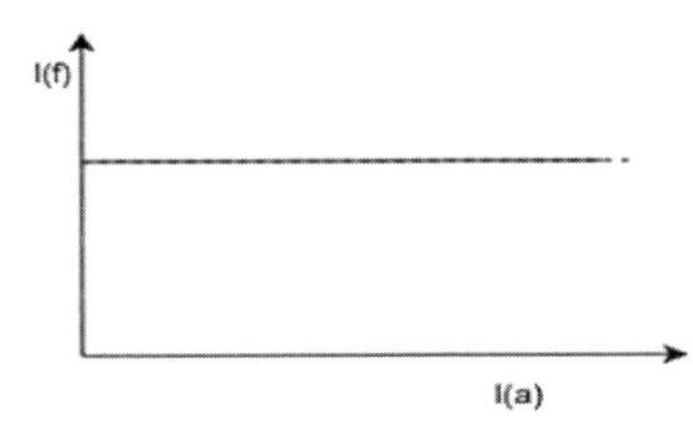


a) Separately excited dc generator, shunt generator

b) Shunt generator, separately excited dc generator

c) Differentially compound dc generator, separately excited

d) Series dc generator, shunt generator

10) The voltage drop in terminal voltage from no-load to full load in a shunt generator can be compensated using

a) Aiding series field

b) Long-shunt, differential field

c) Aiding shunt field

d) Any of the measures

# 8

# Speed Control

**Aim:** To perform speed control of D.C. shunt motor by

1. Armature resistance control method.
2. Field (flux) control method.

## Apparatus required

| S.No. | Device | Rating / Range | Type | Quantity |
|---|---|---|---|---|
| 1. | D.C. Shunt motor | | | 1 |
| 2. | Ammeter | (0-5) A | MC | 1 |
| 3. | Ammeter | (0-10) A | MC | 1 |
| 4. | Voltmeter | (0-250) V | MC | 2 |
| 5. | Tachometer | (0-2000) RPM | Digital | 1 |
| 6. | Rheostats | (100 Ω/2.5A) | | 2 |

## Theory

The speed and torque of a dc motor is given by

$$N = \frac{K_1 \left(V_a - I_a R_a\right)}{\phi} \left(r.p.m.\right) \qquad (1)$$

And $T \quad K \; I$ (2)

Where $\mathbf{K_1}$ and $\mathbf{K_2}$ are constants of the dc motor.

$R_a$= armature resistance in Ω,

$I_a$= armature current in ampere,

$V_a$= armature voltage in volts,

$N$= speed in r.p.m,

$T$= torque,

As per the above, Torque and speed relation, the speed of dc shunt motor can be varied by changing,

(1) The applied voltage to armature.

(2) The flux.

This method is based on the fact that by varying the voltage available across the armature, the back EMF and hence the speed of the motor can be changed. This is done by inserting a variable resistance $R_C$ (known as controller resistance) in series with the armature as shown in Fig. (a).

$$N\alpha \left[V - I_a\left(R_a + R_C\right)\right] \quad (3)$$

Where $R_C$ = controller resistance

Due to voltage drop in the controller resistance, the back EMF ($E_b$) is decreased. Since N α $E_b$, the speed of the motor is reduced. The highest speed obtainable is that corresponding to $R_C = 0$ i.e., normal speed. Hence, this method can only provide speeds below the normal speed

It is based on the fact that by varying the flux ϕ, the motor speed (N α 1/ ϕ) can be changed and hence the name flux control method.

In this method, a variable resistance (known as shunt field rheostat) is placed in series with shunt field winding as shown in Fig. (b).

The shunt field rheostat reduces the shunt field current $I_f$ and hence the flux ϕ. Therefore, we can only raise the speed of the motor above the normal speed. Generally, this method permits to increase the speed in the ratio 3:1. Wider speed ranges tend to produce instability and poor commutation.

**Circuit diagram**

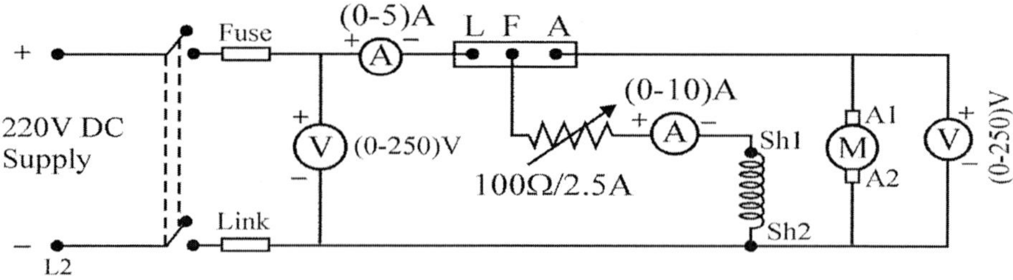


Field Flux Control Method

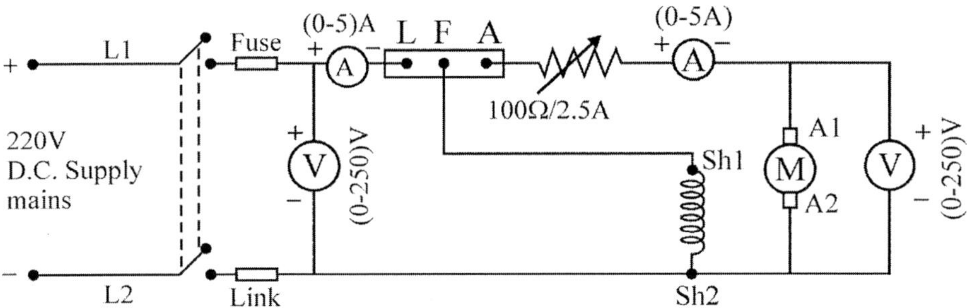


Armature Resistance Control Method

## Procedure

1. Make connections as shown in the figure.
2. Set up the armature circuit resistance in the maximum position and keep the field regulator of the motor at the minimum position i.e.to see that minimum resistance is connected in series with the field at the time starting.
3. Switch on the DC supply and start the motor at no-load.
4. Use a tachometer for measuring the speed (N) of the machine and thus note the speed at various armature resistance positions.
5. For field control, apply the rated voltage across the terminals and decreases the field current in the steps to change the speed of machine.
6. Repeat the above procedure and fill up the table.

## Observation table

## Armature resistance control method

| S.No. | Supply voltage (volts) | Supply current (Amp.) | Armature current ($I_a$) (Amp.) | armature voltage ($V_a$) (volts) | Speed (RPM) |
|---|---|---|---|---|---|
| 1. | | | | | |
| 2. | | | | | |
| 3. | | | | | |
| 4. | | | | | |
| 5. | | | | | |
| 6. | | | | | |
| 7. | | | | | |

## Field flux control method

| S.No. | Supply voltage (volts) | Supply current (Amp.) | Field current ($I_f$) (Amp.) | Constant armature voltage ($V_a$) (volts) | Speed (RPM) |
|---|---|---|---|---|---|
| 1. | | | | | |
| 2. | | | | | |
| 3. | | | | | |
| 4. | | | | | |
| 5. | | | | | |
| 6. | | | | | |
| 7. | | | | | |
| 8. | | | | | |

## Calculation

**Result:** Draw curve showing

1. Speed (N) Vs Back EMF. ($E_b$) with Field current constant ($I_f$).
2. Speed (N) Vs Field current ($I_f$) with Armature voltage constant ($V_a$).

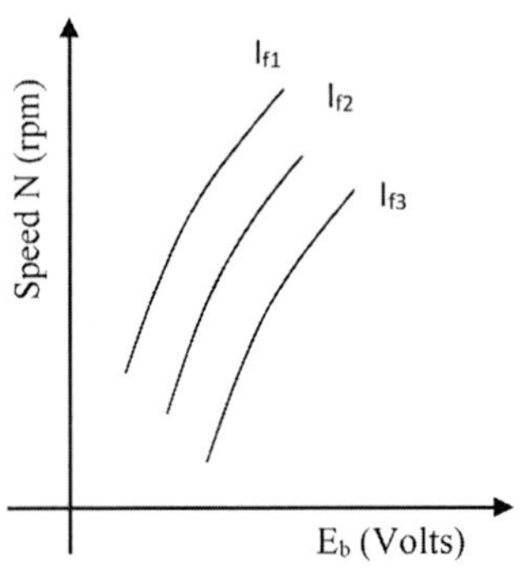


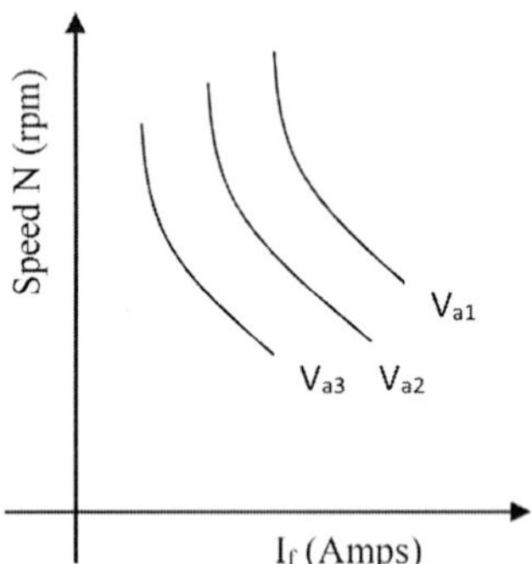


## Discussion

## Precautions

1. Field Rheostat should be kept in the minimum resistance position at the time of starting and stopping the motor.
2. Armature Rheostat should be kept in the maximum resistance position at the time of starting and stopping the motor.

## Viva-Questions

1) In the flux control method for controlling the speed of d.c. shunt motor, the speed control is not possible

a) Below normal rated speed
b) Above normal rated speed
c) Both (a) and (b)
d) None of these

2) In armature voltage control method or rheostatic control method for speed control in d.c. shunt motor, a variable resistance is

a) Added in series with the armature
b) Added in series with the field
c) Connected across armature
d) Both in series with armature and field

3) Which method is suitable for the speed control, below and above the normal rated speed of d.c. shunt motor?

a) Flux control method
b) Rheostatic control method
c) Voltage control method
d) All of these

4) If a resistance is added in series with the field winding of d.c. shunt motor, then

a) Both speed and torque decreases
b) Both speed and torque increases
c) Speed decreases, torque increases
d) Speed increases, torque decreases

5) Speed of d.c. shunt motors are controlled by

a) Flux control method
b) Rheostatic control method
c) Voltage control method
d) All of these

6) With the increase in temperature, the speed of series and shunt motor will

a) Increase, decrease
b) Decrease, increase
c) Increase, increase
d) Decrease, decrease

7) If the speed of a DC shunt motor is increased, the back EMF of the motor will ___________

a) Increase
b) Decrease
c) Remain same
d) Become zero

8) If the speed of a DC shunt motor is increased, the back EMF of the motor will___________

a) Increase
b) Decrease
c) Remain same
d) Become zero

9) Speed regulation of a DC shunt motor is equal to 10%, at no-load speed of 1400 rpm. What is the full load speed?

a) 1233 rpm
b) 1273 rpm
c) 1173 rpm
d) 1123 rpm

10) A series wound motor is also called as universal motor because

a) It will run equally well using either an ac or a dc voltage source
b) It will run well below and above rated speed
c) It can be used in all kinds of applications
d) None of these

# 9

# Swinburne's Test

**Aim:** To determine the Efficiency Vs Load characteristics (External characteristics) of D.C. shunt motor Swinburne's method.

**Apparatus required:**

| S.No. | Device | Type | Range | Quantity |
|---|---|---|---|---|
| 1. | D.C. Shunt motor | | | 1 |
| 2. | Ammeter | Moving Coil | (0-5)A | 2 |
| 4. | Voltmeter | Moving Coil | (0-250) V | 2 |
| 5. | Tachometer | Digital | (0-2000) RPM | 1 |
| 6. | Rheostats | | (145 Ω/2.8A) | 1 |
| 7. | Multimeter | | | 1 |

## Theory

This method is an indirect method of testing a DC machine. It is named after Sir James Swinburne. **Swinburne's test** is the most commonly used and simplest method of testing of shunt and compound wound DC machines which have constant flux. In this test the efficiency of the machine at any load is pre-determined. We can run the machine as a motor or as a generator. The iron and friction losses are determined by measuring the input to the DC machine at no-load. The machine is to be run as a motor at normal voltage and speed. The Swinburne's method includes all the losses in the field circuit and winding which do not occur actually in the machine but the error introduced due to this is quite insignificant and still shows the accurate results. In this method of testing no-load losses are measured separately and eventually we can determine the efficiency.

## Circuit diagram

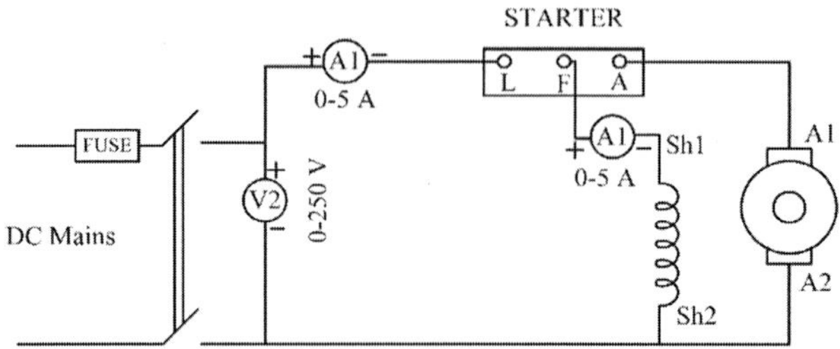


SWINBURNE'S TEST

## Observation table

| S.No. | Motor input voltage (V) | Motor input current ($I_o$) | Motor field current ($I_{Sh}$) | Motor armature current ($I_a$) | Power input at no-load | Armature cu loss at no-load |
|---|---|---|---|---|---|---|
| 1. | | | | | | |
| 2. | | | | | | |
| 3. | | | | | | |
| 4. | | | | | | |
| 5. | | | | | | |

| S.No. | Voltage across armature ($V_a$) | Current through armature ($I_a$) | Armature resistance ($R_a$) | Mean resistance |
|---|---|---|---|---|
| 1. | | | | |
| 2. | | | | |
| 3. | | | | |

## Calculation

No-load Armature current is $I_o - I_{sh}$

Where,

$I_o$ is no-load current

$I_{sh}$ is field current

In case of shunt motor, the field flux remains constant therefore field current also remains constant. Thus, the field copper losses are taken as constant losses. The only variable loss is armature copper loss.

No-load input power is $V^* I_o$

Then Armature copper losses are given as $(I_o - I_{sh})^2 * Ra$

Where,

$R_a$ is the armature resistance.

Constant losses $W_C = V^* I_o - (I_o - I_{sh})^2 * R_a$

Once the constant losses are obtained, the efficiency at any load can be calculated as only armature copper loss will vary according to the load current.

Total losses are $W_C$ + Armature copper losses

Now,

$$\text{Efficiency } \eta = \frac{Output}{Input} = \frac{Input - losses}{Input}$$

$$\eta = \frac{Input - (W_C + Armature\ \ Cu\ \ loss\ \ at\ \ any\ \ load)}{Input}$$

**Result**

Draw graph between efficiency and load at various loadings.

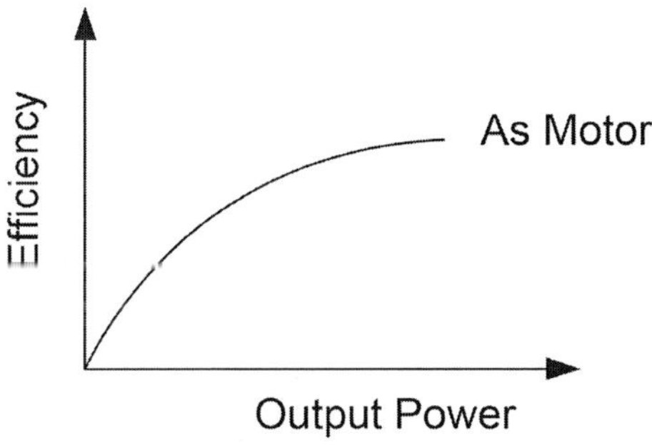


## Advantages of swinburne's Test

The main advantages of this test are:

1) This test is very convenient and economical as it is required very less power from supply to perform the test.
2) Since constant losses are known, efficiency of Swinburne's test can be pre-determined at any load.

## Disadvantages of Swinburne's Test

The main disadvantages of this test are:

1) Iron loss is neglected though there is change in iron loss from no-load to full load due to armature reaction.
2) We cannot be sure about the satisfactory commutation on loaded condition because the test is done on no-load.
3) We can't measure the temperature rise when the machine is loaded. Power losses can vary with the temperature.
4) In DC series motors, the Swinburne's test cannot be done to find its efficiency as it is a no-load test.

## Discussion

## Precautions

1. DC shunt motor should be started and stopped under no-load condition.
2. Field rheostat should be kept in the minimum position.
3. Brake drum should be cooled with water when it is under load

## Viva-Questions

1. Swinburne's test can be carried out on all DC motors.

   a) True    b) False

2. Which of the following test will be suitable for testing two similar DC series motors of large capacity?

   a) Swinburne's test    b) Hopkinson's test

   c) Field test    d) Brake test

3. Which losses can be identified from Swinburne's test?

   a) No-load core loss

   b) Windage and friction loss

   c) No-load and windage and friction loss

   d) Stray load loss

4. While carrying out Swinburne's test at rated armature voltage motor will run at

   a) Speed equal to rated speed

   b) Speed greater than rated speed

c) Speed less than rated speed

d) Can run anywhere

5. In order to run motor on rated speed while carrying out Swinburne's test we add

a) Resistance in parallel with armature

b) Resistance in series with armature

c) Inductor in series with armature

d) Capacitor in parallel with armature

6. What is the purpose of performing retardation test after Swinburne's test?

a) To find stray load loss

b) To find variable losses

c) To separate out windage and friction losses

d) To find shunt field losses

7. Efficiency calculated by Swinburne's test is

a) Exactly equal
b) Over-estimated

c) Under-estimated
d) Depends on the manual errors

8. Which of the following is not a disadvantage of a Swinburne's test?

a) The stray-load losses can't be determined by this test

b) Steady temperature rise can't be determined

c) Does not give results about satisfactory commutation

d) Machine gets damaged

# 10

# Single-Phase Induction Motor

**Aim:** Determination the **equivalent circuit parameters** of a single-phase induction motor.

1. No-load test.
2. Blocked rotor test.

## Apparatus required

| S.No. | Device | Rating | Type | Quantity |
|---|---|---|---|---|
| 1. | Autotransformer | (0-270) V/2.16 kVA | | 1 |
| 2. | 1-phase induction motor | 0-250 V | | 1 |
| 4. | Voltmeter | (0-250) V | MI | 2 |
| 5. | Ammeter | (0-5) A | MI | 1 |
| 6. | Wattmeter | | | 1 |
| 7. | Handle | | | 1 |

## Theory

The performance of a single-phase induction motor is generally studied either by double revolving field theory or by cross field theory, the former being simple and gives a clear physical understanding. In double revolving field theory, the pulsating field mmf vector (F) has been assumed to be the sum of two oppositely rotating mmf vectors $F_f$ (forward) and $F_b$ (backward) of equal magnitudes. Mathematically F = F/2[cos (wt-θ) + cos (wt+θ)]

$F = F_{f+} F_b$ Where, $F = NI$

The product of number of turns in main winding current. Each component of field vector produces the corresponding torque $T_f$ and $T_b$, being equal and opposite at start. Once the motor is started using the starting MMF, it continues the rotation, the net torque being given by ($T_f$ - $T_b$). The equivalent circuit of a single-phase induction motor is also drawn in figure (a) at any slips where the forward and backward impedance $Z_f$ and $Z_b$ are given by,

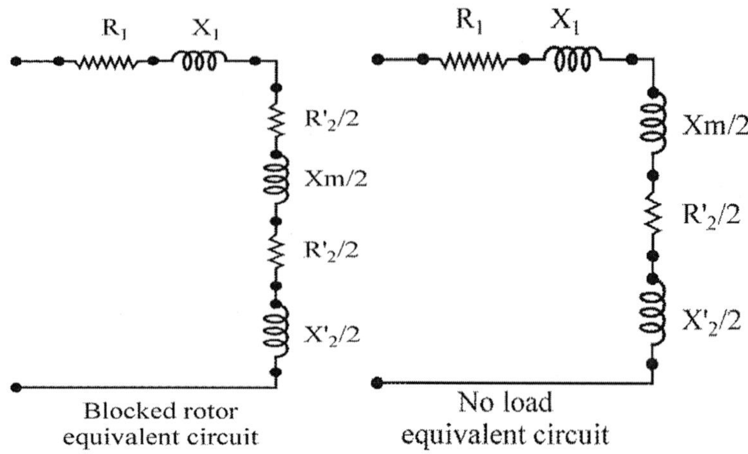


Blocked rotor equivalent circuit

No load equivalent circuit

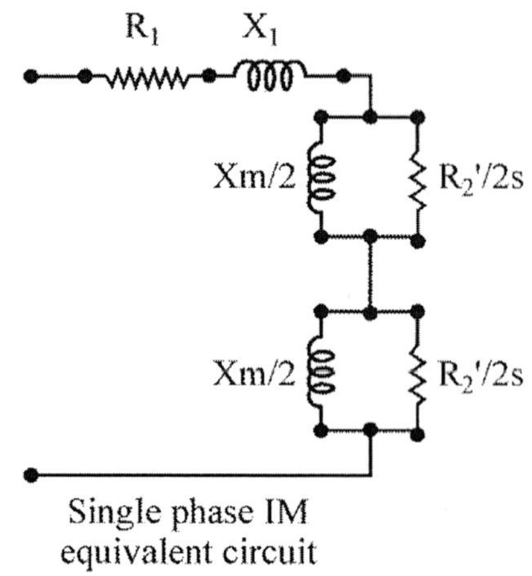


Single phase IM equivalent circuit

$$Z_f = \frac{\left(\frac{R_2}{2s} + \frac{jX_2}{2}\right) \times \frac{jX_m}{2}}{\left(\frac{R_2}{2s} + \frac{jX_2}{2} + \frac{jX_m}{2}\right)}$$

$$Z_b = \frac{\left(\frac{R_2}{2(2-s)} + \frac{jX_2}{2}\right) \times \frac{jX_m}{2}}{\left(\frac{R_2}{2s} + \frac{jX_2}{2} + \frac{jX_m}{2}\right)}$$

## Circuit diagram

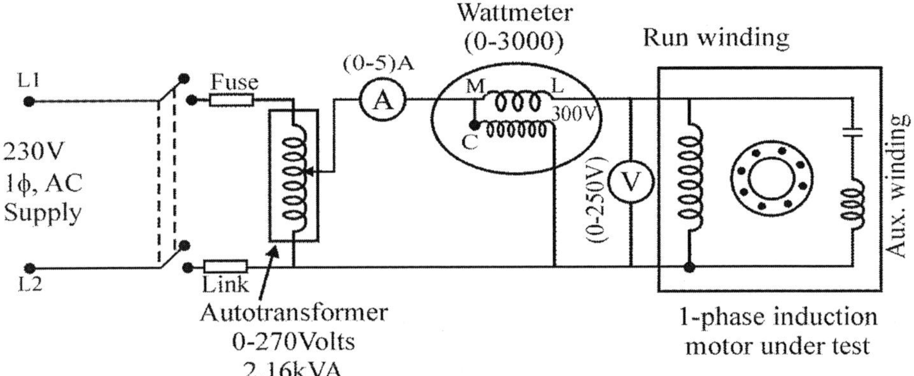


## Procedure

### No-load Test

1. Connect the circuit as per the given circuit diagram.
2. Ensure that the motor is unloaded and the variance is set at zero.
3. Switch ON the supply and increase the voltage gradually, till the rated voltage of the motor. Thus the motor runs at rated speed under no-load.
4. Record the reading of all the meters, connected in the circuit.
5. Switch OFF the AC supply to stop the motor.

### Blocked Rotor Test

1. Read just the variance at zero position.
2. Change the range of all the instruments for block rotor test as suggest in the discussion on the circuit diagram.
3. Block rotor either by tightening the belt firmly or by hand.
4. Switch on the ac supply and apply reduced voltage, so that the input current drawn by the motor under blocked rotor condition is equal to the full load current of the motor.
5. Record the reading of all the meters, connected in the circuit.
6. Switch OFF the AC supply to stop the motor.

## Observation table

### No-load Test

At no-load, s =0. The equivalent circuit thus reduces to as shown in figure (b).

| S.No. | Wattmeter reading ($W_0$) | Voltmeter reading($V_0$) | Ammeter reading($I_0$) |
|---|---|---|---|
| 1. | | | |
| 2. | | | |

### Blocked Rotor Test

When the rotor is blocked and stator is supplied with full load current with reduced voltage the equivalent circuit is as shown in figure (c)

| S.No. | Wattmeter reading ($W_{br}$) | Voltmeter reading ($V_{br}$) | Ammeter reading ($I_{br}$) |
|---|---|---|---|
| 1. | | | |
| 2. | | | |

### Calculation

From blocked rotor test

$\phi_{br} = cos^{-1}(W_{br}/V_{br}*I_{br})$

$R_{eq} = R_1 + R_2 = (V_{br}/I_{br}) *cos\ \phi_{br}$

$X_{eq} = X_1 + X_2 = (V_{br}/I_{br}) *sin\ \phi_{br}$

Assume $X1 = X_2 = X_{eq}/2$

Find $X_m$ from no-load test equivalent circuit by using $R_1$, $R2$, $X_1 + X_2$ as above.

### Result

Using the above calculated parameters and basic network theorems calculated performance indicators (efficiency, power factor, stator current etc.) at various loads.

### Discussion

### Viva –Questions

1) A capacitor start single-phase induction motor will usually have a power factor of

a) unity
b) 0.8 leading
c) 0.6 leading
d) 0.6 lagging.

2) A capacitor start, capacitor run single-phase induction motor is basically a

a) ac series motor
b) dc series motor
c) 2 phase induction motor
d) 3 phase induction motor.

3) The starting torque of a capacitor start motor is

a) zero
b) low
c) same as rated torque
d) More than rated torque.

4) The torque developed by a split phase motor is proportional to

a) Sine of angle between $l_m$ and $l_s$
b) Cosine of angle between $l_m$ and $I_s$
c) Main winding current, $I_m$
d) Auxiliary winding current,$I_s$

5) A capacitor start single-phase induction motor is switched on the supply with its capacitor replaced by an inductor of equivalent reactance value. It will

a) not start
b) start and run
c) start and then stall
d) none of the above.

6) The starting capacitor of a single-phase motor is

a) Electrolytic capacitor
b) Ceramic capacitor
c) Paper capacitor
d) None of the above.

7) Which of the following is the most economical method of starting a single-phase motor ?

a) Resistance start method
b) Inductance start method
c) Capacitance start method
d) Split-phase method.

8) The number of turns in the starting winding of a capacitor start motor as compared to that for split phase motor is

a) same
b) more
c) less
d) none of the above.

9) In a split phase motor, the ratio of number of turns for starting winding to that for running winding is

a) 2.0
b) more than 1
c) 1.0
d) less than 1

10) A single-phase motor generally used for small air compressor is

a) capacitor start capacitor run motor
b) reluctance motor
c) universal motor
d) shaded pole motor

# 11

# 3-Phase Induction Motor

**Aim:** Perform an experiment to determine the parameters of a three-phase induction machine.

## Apparatus Required

| S.No. | Device | Type | Range | Quantity |
|---|---|---|---|---|
| 1 | Wattmeter | ED type | 0 – 750 W | 02 |
| 2 | Ammeter | MI | 0 – 5 A | 01 |
| 3 | Voltmeter | MI | 0 - 500 V | 01 |
| 4 | 3-phase autotransformer | | 0 – 400 V | 01 |

## Theory

As the circuit modal of an induction motor is similar to that of a transformer, the parameter of the modal can be obtained by no-load test and blocked rotor test.

### No-load test

In this test, the motor is to run at no-load at rated voltage and frequency. The applied voltage and current and power input is measured. Power input at no-load ($P_0$) provides only the losses that are rotational losses.

$$P_o = P_{cu}\ (stator\ Cu\ loss) + P_i\ (Iron/core\ loss) + P_{wf}\ (friction + windage\ loss)$$

### Blocked rotor test

It is used to determine the series parameters of induction motor. It corresponds to making S (slip) = 1. This means that rotor must be stationary during the test which requires that rotor must be blocked mechanically while stator is excited with appropriately reduced voltage.

## Observation Table

### No-load test

| S.No. | $V_o$ | $I_o$ | $W_{o1}$ | $W_{o2}$ |
|---|---|---|---|---|
| | | | | |

### Block rotor test

| S.No. | $V_{sc}$ | $I_{sc}$ | $W_{sc1}$ | $W_{sc2}$ |
|---|---|---|---|---|
| | | | | |

## Circuit Diagram

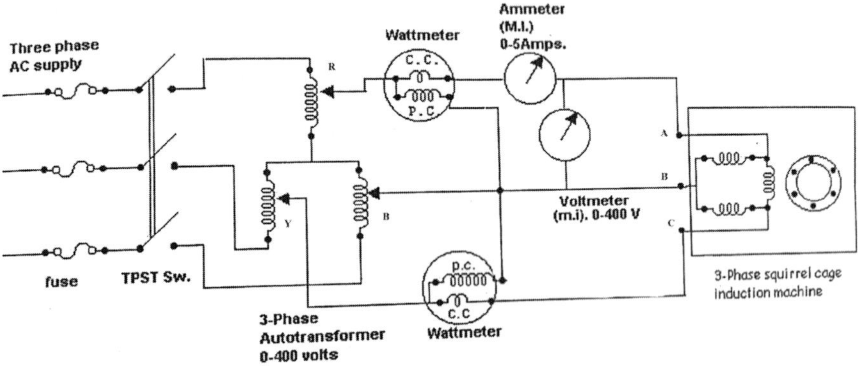


## Calculations

### No-load test

$$Z_o = \frac{V_o}{I_o}$$

$$R_o = \frac{P_o}{I_o^2}$$

$$X_o = \sqrt{\left(Z_o^2 - R_o^2\right)}$$

$$X_m = X_o + X_1$$

**Blocked rotor test**

$$R_{br} = \frac{P_{br}}{I_{br}}$$

$$Z_{br} = \frac{V_{br}}{I_{br}}$$

$$X_{br} = (Z_{br} - R_{br})$$

$$X_1 = X_2 = \frac{X_{br}}{2}$$

$$R_2 = R_{br} - R_1 \times \frac{X_2^{\,2}}{X_1^{\,2}}$$

**Results**

**No-load test**

$Cos\Theta_o =$

$I_w =$

$I_m =$

$R_o =$

$X_o =$

**Block rotor test**

$R_{br} =$

$Z_{br} =$

$X_{br} =$

**Precautions**

1.) The connections should be tight.

2.) Ammeter ,voltmeter, Wattmeter should be of proper range.

3.) Thick copper wires need to be used for connections which have lower resistance that can be neglected.

**Viva-Questions**

1. What machine parameters can be obtained from No-Load test?
2. What is the power factor of the machine? Comment on its value.

3. What should be the no-load current of an induction motor?
4. Even though there is no-load, why wattmeter reading is not zero?
5. Comment on the slip of the machine when operated at rated voltage.
6. How to obtain the no-load input power to an induction motor when two-wattmeter method of measuring power used?
7. Can a three phase induction motor be started from a single-phase supply?
8. No-load test is conducted at (a)rated current, (b)rated voltage, (c)high voltage, (d)high current
9. What is the nameplate reading on the machine? What inferences can be drawn from it?
10. What is the real and reactive power consumed in this test?
11. What are the different losses that are present in an induction machine?
12. Which loss in the machine is significant in no-load test and why?

# 12

# Load Test on Three Phase Induction Motor

## Aim

(a) Perform load test on 3-phase induction motor.

(b) Compute Torque, Output power, Input power, Efficiency, Input power factor and Slip for every load setting and to determine how speed, efficiency, power factor, stator current torque, and slip of an induction motor vary with load.

## Apparatus Required

| S.No. | Devices | Type | Range | Quantity |
|---|---|---|---|---|
| 1 | Ammeter | MI | 5 A | 1 |
| 2 | Voltmeter | MI | 500 V | 1 |
| 3 | Wattmeter | ED | 2.5/5 A, 150/300/600 V | 2 |
| 4 | DOL Starter | Push button type | For 2/3 HP motors | 1 |

## Machine Required

A.C. Motor 2/3 H.P. 3 Phase 415 V 3.6/4.8 A 1440 RPM with Drum Brake Loading Arrangement.

## Theory

The load test on induction motor is performed to compute its complete performance i.e. torque, slip, efficiency, power factor etc. During this test, the motor is operated at rated voltage and frequency and normally loaded mechanically by brake and pulley arrangement from the observed data, the performance can be calculated, following the steps given below.

Slip: The speed of rotor, Nr droops slightly as the load on the motor is increased. The synchronous speed, Ns of the rotating magnetic field is calculated, based on the number of poles, P and the supply frequency, f i.e.

Synchronous Speed, $N_S = \frac{120f}{P} r.p.m$

then Slip, S= $\frac{N_S - N_r}{N_S} * 100\ percent$

Normally, the range of slip at full load is from 2 to 5 percent.

**Torque:** Mechanical loading is the most common type of method employed in laboratories, A brake drum is coupled to the shaft of the motor and the load is applied by tightening the belt, provided on the brake drum. The net force exerted at the brake drum in kg is obtained from the readings S1 and S2 of the spring balances i.e.

Output = Torque x Speed

Thus as the speed of motor does not vary appreciably with load torque will increase with increasing load.

Net force exerted, W = $(S_1 - S_2)$kg

Then, load torque, $T = W * \frac{d}{2}\ kg - m$

$$T = W * \frac{d}{2} * 9.8\ Nw - m$$

Where, d – effective diameter of the brake drum in meters.

**Output Power, P0:** The output power in watts developed by the motor is given by,

Output power, $P_o = \frac{2\pi NT}{60}$

Where, N is the speed of the motor in r. p. m.

**Input Power:** Input power is measured by the two wattmeters', properly connected in the circuit i.e.

Input Power = $(W_1 + W_2)$ watts

Where, W1 and W2 are the readings of the two wattmeter's.

**Power Factor:** Power factor of induction motor on NO-LOAD is very low because of the high value of magnetizing current. With the increase in load the power factor increases because the power component of the current is increased. Low power factor operation is one of the disadvantages of induction motor. An induction motor draws heavy amount of magnetizing current due to presence of air gap between the stator and rotor. Thus, to reduce the magnetizing current in induction motor the air-gap is kept as small is possible.

Input power factor can also be calculated from the readings of two wattmeter's for balanced load. If ϕ is the power factor angle, then

$$\tan\ \phi = \sqrt{3}\frac{W_1 - W_2}{W_1 + W_2}$$

Knowing the power factor angle ϕ from the above, power factor cos ϕ can be calculated. It may be noted clearly at this stage, that the power factor of the induction motor is very low at no-load, hardly 0.1 to 0.25 lagging. As such, one of the wattmeter will record a negative reading, till the power factor is less than 0.5, which may be measured by revering the connection of either the current coil or pressure coil of this wattmeter.

## Efficiency

Percentage efficiency of the motor, $\eta = \frac{Output\ power}{Input\ power} * 100$

Full load efficiency of 3 phase induction motor lies in the range of 72 % (for small motors) to 82 % (for very large motors).

**Speed:** When the induction motor is on NO-LOAD speed is slightly below the synchronous speed. The current due to induced emf in the rotor winding is responsible for production of torque required at NO-LOAD. As the load is increased the rotor speed is slightly reduced. The emf induced in the rotor causes the current increased to produce higher torque, until the torque developed is equal to torque required by load on motor.

## Circuit Diagram

The circuit diagram of load test on 3 phase squirrel cage induction motor. Instruments connected in the circuit serve the function indicated against each.

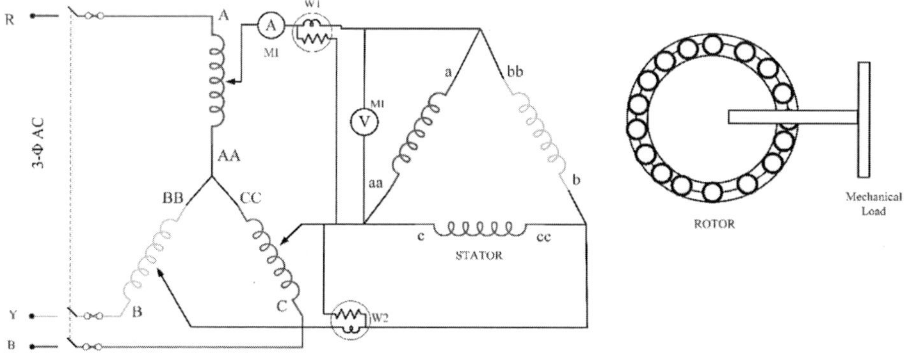

## Procedure

1. Connect the circuit as shown in above figure.
2. Ensure that the motor is unloaded and the variac is set at zero output voltage.
3. Switch-on 3 phase ac mains and start the motor at reduced applied voltage. Increase the applied voltage, till its rated value.
4. Observe the direction of rotation of the motor. In case, it is reverse, change the phase sequence of the applied voltage.
5. Take-down the readings of all the meters and the speed under no-load running.
6. Increase the load on the motor gradually by turning of the hand wheels, thus tighten the belt. Record the readings of all the meters and the speed at every setting of the load. Observation may be continued up to the full load current rating of the motor.
7. Reduce the load on the motor and finally unload it completely.
8. Switch-off the supply to stop the motor.
9. Note-down the effective diameter of the brake drum.

## Observations

| S.No. | Line voltage | Input current | W1 | W2 | S1 | S2 | Speed |
|---|---|---|---|---|---|---|---|
| | | | | | | | |
| | | | | | | | |
| | | | | | | | |
| | | | | | | | |

## Calculation

| S.No. | Current | Input power | Torque | Output power | Slip | Power factor | Efficiency |
|---|---|---|---|---|---|---|---|
| | | | | | | | |
| | | | | | | | |
| | | | | | | | |
| | | | | | | | |
| | | | | | | | |

## Result

Load test on 3 phase Induction Motor has been performed and Input power, Torque, Output power, Slip, Power factor, Efficiency has been obtained at different loading conditions.

## Precautions

While loading the induction motor by brakes, check whether cooling water is circulated in the drum. Before starting the motor, loosen the strap and then tighten it gradually when the motor has picked up speed.

## Viva-Questions

1. A three-phase slip ring induction motor is fed from the rotor side with the stator winding short-circuited. The frequency of the current flowing in the short-circuited stator is

   a) Slip frequency

   b) Supply frequency

   c) The frequency corresponding to rotor speed

   d) Zero

2. An 8-pole, 3-phase, 50 Hz induction motor is operating at a speed of 720 rpm. The frequency of the rotor current of the motor in Hz is

   a) 2 b) 4 c) 3 d) 1

3. Calculate the phase angle of the sinusoidal waveform $z(t)=78\sin(456\pi t+2\pi\div78)$.

   a) $\pi\div39$ b) $2\pi\div5$

   c) $\pi\div74$ d) $2\pi\div4$

4. A 50 Hz, 4poles, a single-phase induction motor is rotating in the clockwise direction at a speed of 1425 rpm. The slip of motor in the direction of rotation & opposite direction of the motor will be respectively.

   a) 0.05, 0.95 b) 0.04, 1.96

   c) 0.05, 1.95 d) 0.05, 0.02

5. The frame of an induction motor is made of

a) Aluminum
b) Silicon steel
c) Cast iron
d) Stainless steel

6. A 3-phase induction motor runs at almost 1000 rpm at no-load and 950 rpm at full load when supplied with power from a 50 Hz, 3-phase supply. What is the corresponding speed of the rotor field with respect to the rotor?

a) 30 revolution per minute
b) 40 revolution per minute
c) 60 revolution per minute
d) 50 revolution per minute

7. In an induction motor, when the number of stator slots is not equal to an integral number of rotor slots

a) There may be a discontinuity in torque slip characteristics
b) A high starting torque will be available
c) The machine performs better
d) The machine may fail to start

8. When the rotor of 3 phase induction motor is blocked, the value of slip will be

(a) zero.
(b) 0.1.
(c) 0.5.
(d) unity

# 13

# To Draw V Curves of Synchronous Motor

**Aim:** To study the effect of variation of field current upon the stator current and power factor with synchronous motor running at no-load, hence to draw V and inverted V curves of the motor.

## Apparatus Required

| S.No. | | Device | Type | Range | Quantity |
|---|---|---|---|---|---|
| 1 | For DC Generator | Voltmeter | MC | 0-300 V | 1 |
| 2 | | Ammeter | MC | 0-10 A | 1 |
| 3 | | Field Rheostat | | 1.4 A/ 230 Ω | 1 |
| 4 | For Synchronous Motor | Voltmeter | MI | 0-600 V | 1 |
| 5 | | Ammeter | MI | 0-5 A | 1 |
| 6 | | PF meter | | | 1 |
| 7 | | Direct Online Starter | | | 1 |
| 8 | | Excitation Switch | | | 1 |
| 9 | For Excitor | Voltmeter | MC | 0-300 V | 1 |
| 10 | | Ammeter | MC | 0-2.5 A | 1 |
| 11 | | DP, MCB | | 2 Amp | |

## Theory

With constant mechanical load on the synchronous motor, the variation of field current changes the armature current drawn by the motor and also its operating power factor. As such, the behavior of the synchronous motor is described below under three different modes of field excitation.

## Normal Excitation

The armature current is minimum at a particular value of field current, which is called the normal field excitation. The operation power factor of the motor is unit at this excitation and thus the motor is equivalent to a resistive type of load.

## Under Excitation

When the field current is decreased gradually below the normal excitation, the armature current increase and the operating power factor of the motor decreases. The power factor under this condition is lagging. Thus, the synchronous motor draws a lagging current, when it is under excited and is equivalent to an inductive load.

## Over Excitation

When the field current is increased gradually beyond the normal excitation, the armature current again increase and the operating power factor increases. However, the power factor is leading under this condition. Hence, the synchronous motor draws a leading current, when it is over excited and is equivalent to a capacitive load.

If the above variation of field current and the corresponding armature current are plotted for a constant mechanical load, a curve of the shape of 'V' is obtained as shown in figure below. Such a characteristic of synchronous motor is commonly called as 'V' curve of the motor. The characteristic curve plotted between input power factor and the field current for a constant mechanical load on the motor are of the shape of inverted 'V' and are known as inverted 'V' curves.

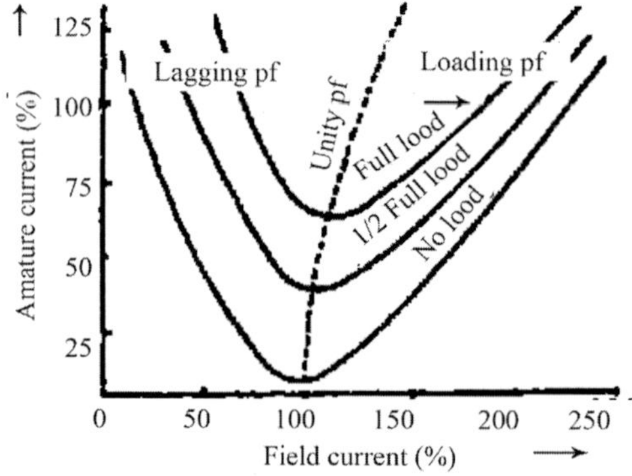


**Fig.1.** V-curve of Synchronous Motor

For increased constant mechanical load on the motor, 'V' curves bodily shift upwards as shown in fig 'A'. The curve joining the minimum current points of various 'V' curves plotted for different mechanical loads is normally called a unity power factor compounding curve.

## Circuit Diagram

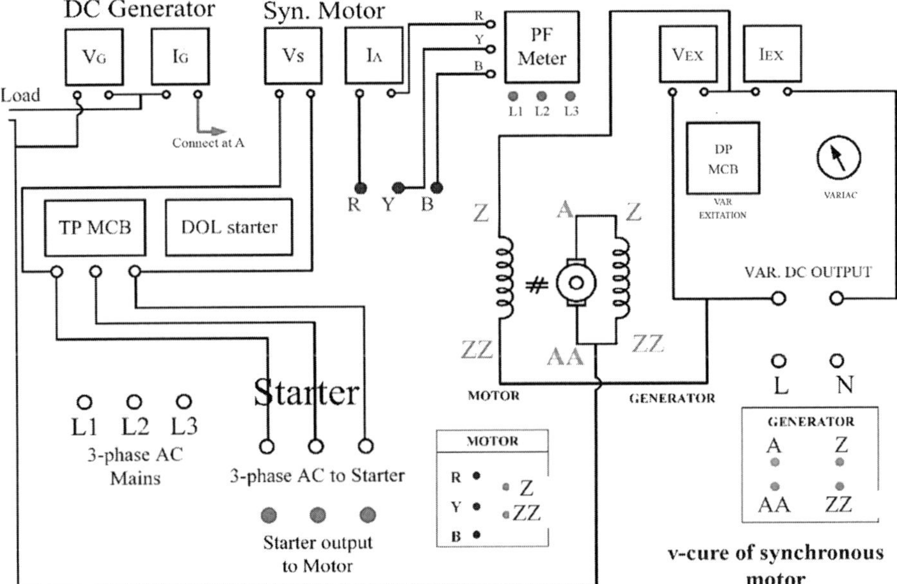


v-cure of synchronous motor

## Procedure

1. Connect the ac three phase mains to terminals marked L1', L2', L3'. Switch ON the MCB.
2. Connect single-phase AC supply on terminals marked L & Ne in order to get variable DC Excitation for field circuit of auto Synchronous Motor. MCB is to be turned ON only when the DC excitation is to be given to the field of Auto Synchronous Motor.
3. Before starting the synchronous motor makes sure that the DC excitation switch is in OFF position.
4. For starting the MG set press green push button of DOL starter in the marked direction (by arrow). The synchronous will start as induction motor. The corresponding current will be shown by ammeter and the power factor lag shown by three phase power factor meters. Note the readings of (Ia). Also note the corresponding terminals voltage of shunt generator.
5. Switch ON the DC Excitation switch. The armature current will slightly decrease. Note this reading.

6. Now gradually move the static excitation controller (Variac) in clockwise direction by a few steps and note the corresponding decrease in armature current (Ia) and field excitation current (If). Vary the excitation till the armature current is minimum. After this point on moving the Knob of Variac, the armature current will increase and note corresponding armature current and field current values. Vary the excitation up till the rated value of synchronous motor. This value will correspond to synchronous motor at no-load.
7. Adjust the voltage of DC generator coupled to synchronous motor to rated value by varying the field rheostat of DC generator (1.2 Amp, 260 Ohms).
8. Now connect load on DC Generator at terminals L1 & L2 by means of single-phase loading rheostat/Lamp Load. Now draw the V curves at one fourth load, half load and three fourth loads by gradually loading in steps.
9. Now for various load settings on DC generator repeat steps No. 2 to 5. The load setting must be maintained constantly for drawing V curves at that particular load.

## Observation Table

| S.No. | V | $I_f$ | $I_a$ | $V_{dc}$ | $I_{dc}$ | PF |
|---|---|---|---|---|---|---|
| | | | | | | |
| | | | | | | |
| | | | | | | |

## Result

The V curve and inverted V curve is obtained.

## Discussion

## Viva-Questions

1. If graph of armature current drawn by synchronous motor is against field current is plotted then the resulted graph is known as

   a) V-curves b) Inverted V-curves

   c) Both (a) & (b) d) None of the above

2. The plot of power factor against the field current of synchronous motor is known as
   a) V-curves
   b) Inverted V-curves
3. The magnetization current drawn from an AC supply a synchronous motor is used to
   a) set up flux in magnetic circuit of device
   b) compensate core losses
   c) set up magnetizing armature reaction
   d) all of the mentioned
4. A 3 phase synchronous motor is working at normal excitation, then the flux deficient in circuit is
   a) given by armature winding mmf
   b) given by field winding mmf
   c) supplied to armature winding mmf
   d) supplied to field winding mmf
5. Synchronous compensators are
   a) overexcited synchronous motor with no mechanical load
   b) overexcited synchronous motor with mechanical load
   c) under excited synchronous motor with no mechanical load
   d) normal excited synchronous motor with no mechanical load
6. For a power system having induction motor loads, an overexcited synchronous motor is also attached. The induction motor will now operate at
   a) lagging
   b) leading
   c) reduced power factor
   d) increased power factor
7. While running, a synchronous motor is compelled to run at synchronous speed because of
   a) Damper winding in its pole faces
   b) Magnetic locking between stator and rotor poles
   c) Induced e.m.f. in rotor field winding by stator flux
   d) Compulsion due to Lenz›s law
8. The speed regulation of a synchronous motor is always
   a) 1%
   b) 0.5%
   c) Positive
   d) Zero

# 14

# 3-Point and 4-Point Starter

**Aim:** Study of 3 point and 4-point starter.

## Theory

A **3-point starter** in simple words is a device that helps in the starting and running of a shunt field or compound excited dc motor. Now the question is why these types of dc motors requires the assistance of the starter in the first case. The only explanation to that is given by the presence of back emf $E_b$, which plays a critical role in governing the operation of the motor. The back emf, develops as the motor armature starts to rotate in presence of the magnetic field, by generating action and counters the supply voltage. This also essentially means that the back emf at the starting is zero, and develops gradually as the motor gathers speed.

The general motor emf equation

$$E = E_b + I_a R_a \tag{1}$$

at starting is modified to $E = I_a.R_a$ as at starting $E_b = 0$.

$$I_a = \frac{E}{R_a} \tag{2}$$

Thus we can well understand from the above equation that the current will be dangerously high at starting (as armature resistance $R_a$ is small) and hence its important that we make use of a device like the **3 point starter** to limit the starting current to an allowable lower value.

Let us now look into the construction and **working of three point starter** to understand how the starting current is restricted to the desired value. For that let's consider the diagram given below showing all essential parts of the three-point starter.

## Construction of 3-point starter

Construction wise a starter is a variable resistance, integrated into number of sections as shown in the figure beside. The contact points of these sections are called studs and are shown separately as **OFF, 1,2,3,4,5, RUN**. Other than that, there are 3 main points, referred to as

i. 'L' Line terminal. (Connected to positive of supply.)

ii. 'A' Armature terminal. (Connected to the armature winding.)

iii. 'F' Field terminal. (Connected to the field winding.)

And from there it gets the name 3 point starter

Now studying the construction of 3 point starter in further details reveals that, the Point 'L' is connected to an electromagnet called overload release (OLR) as shown in the figure.

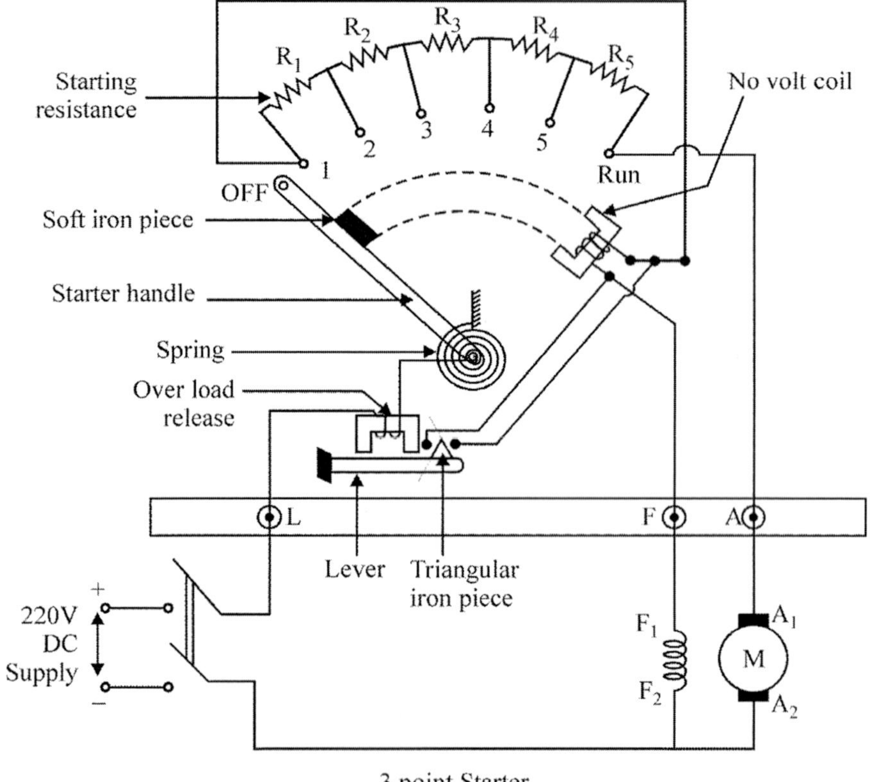


3 point Starter

The other end of 'OLR' is connected to the lower end of conducting lever of starter handle where a spring is also attached with it and the starter handle contains also a soft iron piece housed on it. This handle is free to move to the other side RUN against the force of the spring. This spring brings back the handle to its original OFF position under the influence of its own force. Another parallel path is derived from the stud '1', given to another electromagnet called No Volt Coil (NVC) which is further connected to terminal 'F'. The starting resistance at starting is entirely in series with the armature. The OLR and NVC acts as the two protecting devices of the starter.

## Working of Three Point Starter

Having studied its construction, let us now go into the **working of the 3-point starter**. To start with the handle is in the OFF position when the supply to the d.c. motor is switched on. Then handle is slowly moved against the spring force to make a contact with stud No. 1. At this point, field winding of the shunt or the compound motor gets supply through the parallel path provided to starting resistance, through No Voltage Coil.

While entire starting resistance comes in series with the armature. The high starting armature current thus gets limited as the current equation at this stage becomes

$$I_a = \frac{E}{\left(R_a + R_{st}\right)} \tag{3}$$

As the handle is moved further, it goes on making contact with studs 2, 3, 4 etc., thus gradually cutting off the series resistance from the armature circuit as the motor gathers speed. Finally when the starter handle is in 'RUN' position, the entire starting resistance is eliminated and the motor runs with normal speed. This is because back emf is developed consequently with speed to counter the supply voltage and reduce the armature current. So the external resistance is not required anymore, and is removed for optimum operation. The handle is moved manually from OFF to the RUN position with development of speed. Now the obvious question is once the handle is taken to the RUN position how is it supposed to stay there, as long as motor is running?

To find the answer to this question let us look into the working of No Voltage Coil.

## Working of No Voltage Coil of 3-point starter

The supply to the field winding is derived through No Voltage Coil. So when field current flows, the NVC is magnetized. Now when the handle is in the 'RUN' position, soft iron piece connected to the handle and gets attracted by the magnetic force produced by NVC, because of flow of current through it. The NVC is designed in such a way that it holds the handle in ‹RUN› position against the force of the spring as long as supply is given to the motor. Thus

NVC holds the handle in the ‹RUN› position and hence also called **hold on coil**.

Now when there is any kind of supply failure, the current flow through NVC is affected and it immediately loses its magnetic property and is unable to keep the soft iron piece on the handle, attracted. At this point under the action of the spring force, the handle comes back to OFF position, opening the circuit and thus switching off the motor. So due to the combination of NVC and the spring, the starter handle always comes back to OFF position whenever there is any supply problems. Thus, it also acts as a protective device safeguarding the motor from any kind of abnormality.

## 4-Point Starter

The **4 point starter** like in the case of a 3 point starter also acts as a protective device that helps in safeguarding the armature of the shunt or compound excited dc motor against the high starting current produced in the absence of back emf at starting. The 4 point starter has a lot of constructional and functional similarity to a three point starter, but this special device has an additional point and a coil in its construction, which naturally brings about some difference in its functionality, though the basic operational characteristic, remains the same. Now to go into the details of **operation of 4-point starter**, let's have a look at its constructional diagram, and figure out its point of difference with a 3 point starter.

## Construction and Operation of four-point Starter

A 4-point starter as the name suggests has 4 main operational points, namely

i. 'L' Line terminal. (Connected to positive of supply.)

ii. 'A' Armature terminal. (Connected to the armature winding.)

iii. 'F' Field terminal. (Connected to the field winding.)

Like in the case of the 3 point starter, and in addition to it there is 4. A 4th point N. (Connected to the No Voltage Coil)

The remarkable difference in case of a 4 point starter is that the No Voltage Coil is connected independently across the supply through the fourth terminal called 'N' in addition to the 'L', 'F' and 'A'. As a direct consequence of that, any change in the field supply current does not bring about any difference in the performance of the No Voltage Coil. Thus it must be ensured that No Voltage Coil always produce a force which is strong enough to hold the handle in its ‹RUN› position, against force of the spring, under all the operational conditions. Such a current is adjusted through No Voltage Coil with the help of fixed resistance R connected in series with the NVC using fourth point 'N' as shown in the figure above.

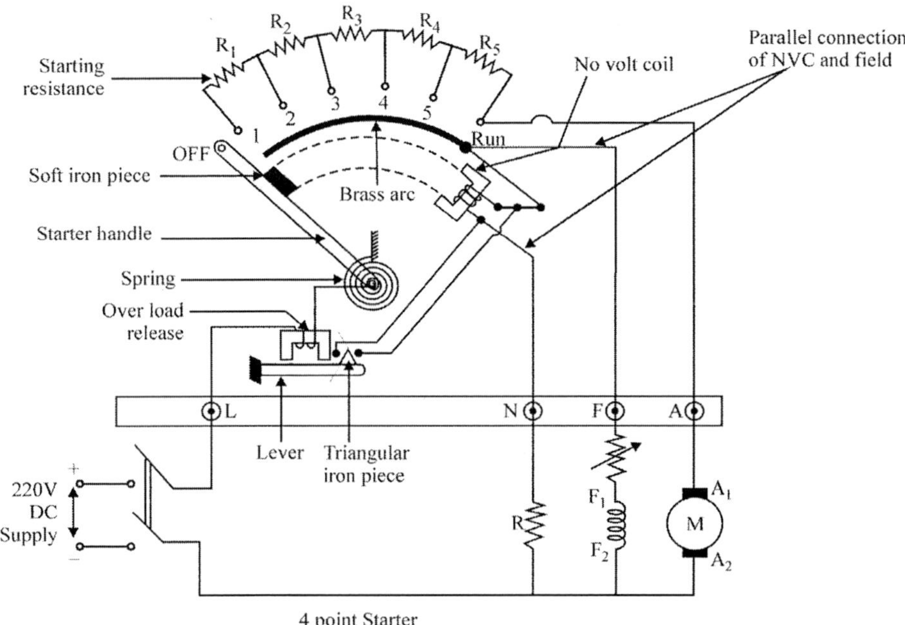

4 point Starter

Apart from this above mentioned fact, the 4 point and 3 point starters are similar in all other ways like possessing is a variable resistance, integrated into number of sections as shown in the figure above. The contact points of these sections are called studs and are shown separately as **OFF, 1, 2, 3, 4, 5, RUN**, over which the handle is free to be manoeuvred manually to regulate the starting current with gathering speed.

Now to understand its way of operating let's have a closer look at the diagram given above. Considering that supply is given and the handle is taken stud No. 1, then the circuit is complete and line current that starts flowing through the starter. In this situation we can see that the current will be divided into 3 parts, flowing through 3 different points.

i. 1 part flows through the starting resistance ($R_1$+ $R_2$+ $R_3$.....) and then to the armature.

ii. 2nd part flowing through the field winding F.

iii. 3rd part flowing through the No Voltage Coil in series with the protective resistance R.

So the point to be noted here is that with this particular arrangement any change in the shunt field circuit does not bring about any change in the No Voltage Coil as the two circuits are independent of each other. This essentially means that the electromagnet pull subjected upon the soft iron bar of the handle by the No Voltage Coil at all points of time should be high enough to keep the handle at its RUN position, or rather prevent the spring force from restoring the handle at its original OFF position, irrespective of how the field rheostat is adjusted.

This marks the operational difference between a 4 point starter and a 3 point starter. As otherwise both are almost similar and are used for limiting the starting current to a shunt field or compound excited dc motor, and thus act as a protective device.

## Result

We have studied 3-point and 4-point starter and find that 4-point starter is more suitable as compared to 3-point starter.

## Discussion

## Viva-Questions

1) Which starter does not provide high speed protection to the d.c. shunt motor?

   a) Three point starter
   b) Four point starter
   c) Two point starter
   d) None of these

2) The basic difference between three-point starter and four-point starter is

   a) No volt coil is connected in series with the field winding in three point starter while connected independently across the supply in four point starter
   b) No volt coil is connected across supply in three point starter while in series with field winding in four point starter.
   c) No volt coil is connected across field winding in three point starter while connected across supply in four point starter.
   d) None of these

3) 3 - point starter is used to start the

a) Series motor
b) Shunt motor
c) Compound motor
d) Only (b) and (c)

4) Why starters are required in a DC motor?

a) Back emf of these motors is zero initially
b) These motors are not self-starting
c) These motors have high starting torque
d) To restrict armature current as there is no back emf at starting

5) The starting resistance of a DC shunt motor is generally ______

a) Low
b) Around 0.5 kΩ
c) Around 5 kΩ
d) Infinitely large

6) What will happen if DC motor is used without starter?

a) Heavy sparking at brushes
b) It'll start smoothly
c) Will not start at all
d) Depends on load

7) Considering a human handed control system for the dc motor speed control, if the resistance wire cut out too slowly, then

a) Starting resistance would burn
b) Field winding would burn
c) Speed will rise steeply
d) Any of the mentioned

8) A starting resistance is inserted at the starting in an induction motor as well as dc motor.

a) Induction motor has to control starting torque whereas in dc motor, it is done to avoid large current
b) To limit starting current in both the machines
c) To limit starting speed
d) All of the mentioned

9) Which of the following express the starting current nature of the dc motor

a)

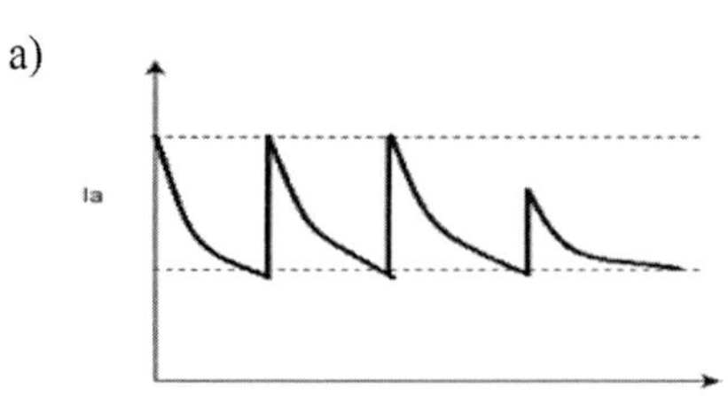


b)

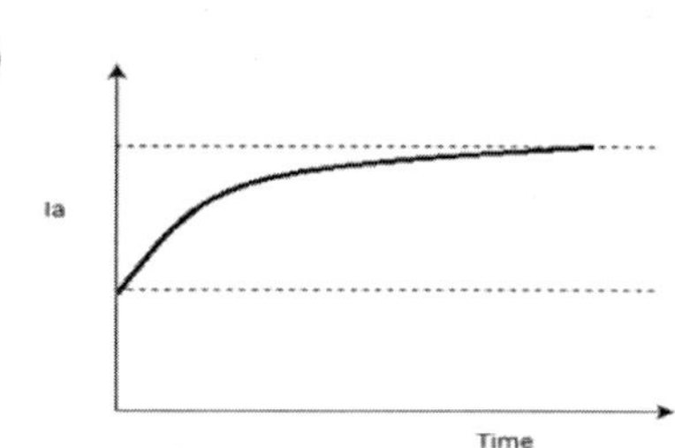


c)

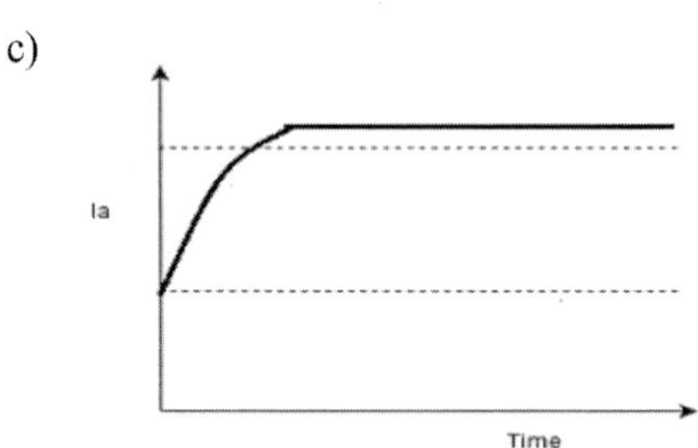


d)

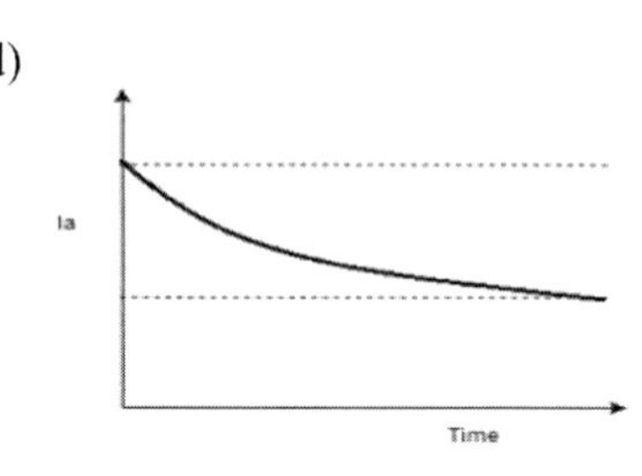


10) How does the speed build up takes place in a dc motor with time?

a)

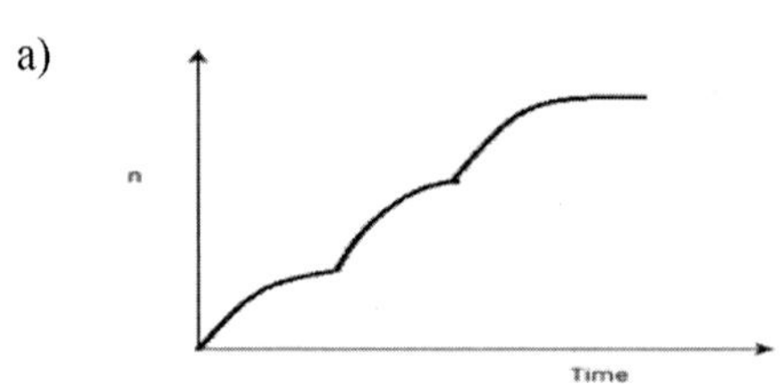


b)

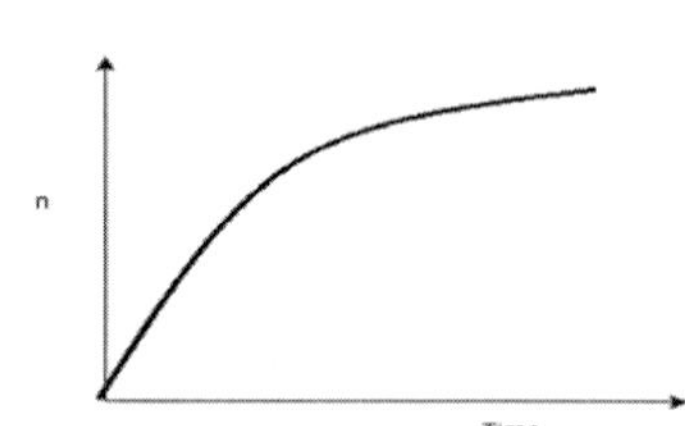


c)

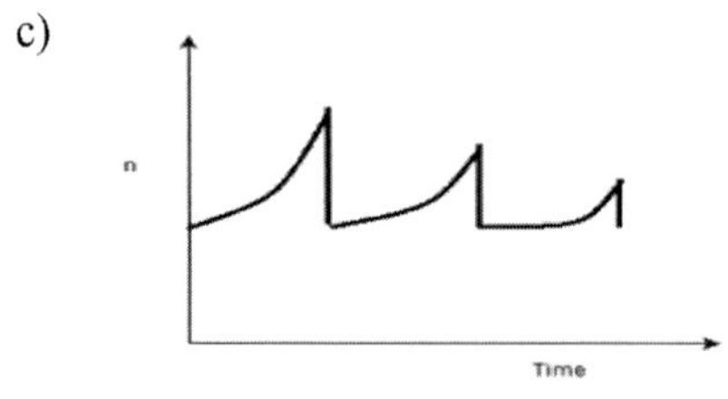


d)

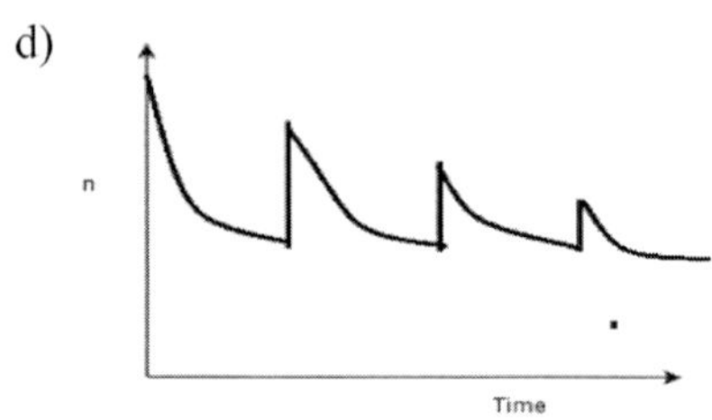

# APPENDIX A

## Objective Questions and Answers on Electrical Machines

## I. Transformers

1. Low voltage windings are placed nearer the core in the case of concentric windings because

   (a) It reduces leakage fluxes

   (b) It reduces insulation requirement

   (c) It reduces eddy current loss

   (d) It reduces hysteresis loss.

2. For voltage above 66 kV, condenser bushings are preferred than non-condenser bushings because

   (a) The radial stresses can be made independent of the radial thickness of the dielectric

   (b) Optimum utilization of the dielectric can be made due to uniform potential distribution which also reduces both the radial and the axial dimensions of the bushings

   (c) The axial stresses can be eliminated

   (d) None of the above.

3. Continuous disc winding is suitable for

   (a) High voltage winding of large transformers

   (b) Low voltage winding of small transformers

   (e) High voltage winding of small transformers

   (d) Low voltage winding of large transformers.

4. Non-loading heat run test on transformers is performed by means of

   (a) SC test

   (b) Half time on SC and half time on OC

   (c) Sumpner's test

   (d) OC test.

5. In a transformer the flus phasor
   (a) Lags the induced e.m.f. by slightly less than 90°
   (b) Leads the induced e.m.f. by slightly less than 90°
   (c) Lags the induced e.m.f. by 90°
   (d) Leads the induced e.m.f. by 90°.
6. In a Scott-connected transformer the number of primary and teaser turns respectively are:
   (a) N, √3 N/2 (b) N, 2/√3 N
   (c) √3 N/2, N (d) N/2, N
7. For off-load tap changing the best method is to use
   (a) Tap changers outside the tank operated by selector switches
   (b) Tap changers outside the tank but with no selector switch
   (c) Tap changers inside the tank operated by external selector switches
   (d) Tap changers inside the tank but with no selector switch.
8. The amount of leakage flux in the transformer windings depends upon:
   (a) The mutual flux (b) The load current
   (c) Turn ratio (d) The applied voltage
9. For small power transformers, it is preferable to use
   (a) Corrugated tanks (b) Radiator tanks
   (c) Tanks with separate coolers (d) Tubed tanks.
10. The transformer windings are tapped in the middle because
   (a) It eliminates axial forces on the windings
   (b) It eliminates radial forces on the windings
   (c) It reduces insulation requirement
   (d) None of the above.

11. Power and distribution transformers are used under which of the load conditions specified below

| Power Transformer | Distribution Transformer |
|---|---|
| (a) All full-load or higher only | Close to 3/4th full-load with minor variations |
| (b) Wide load variation from half full-load to full-load | At any load up to full-load but with minor variation only. |
| (c) Close to full-load with minor variation | Wide load variations from no-load to full load |
| (d) At full-load or higher only | Close to 3/4th full-load with min variations. |

12. The efficiency of a transformer at full-load .85 pf lag is 95%. Its efficiency at fu load 0.85 pf lead will be

(a) More than 95%

(b) 100%

(c) 95%

(d) Less than 95%.

13. For on-load tap changing the best method is to use

(a) Tap changers outside the tank but with no selector switches

(b) Tap changers outside the tank operated by selector switches

(c) Tap changers inside the tank operated by external selector switches

(d) Tap changers inside the tank but with no selector switch.

14. The purpose of providing a conservator on a transformer tank is

(i) To prevent oil from coming in contact with the atmosphere

(ii) To permit breathing and to increase the oil surface exposed to atmosphere

(iii) To permit breathing and yet to reduce the oil surface exposed to atmosphere,

Mark the correct answer below:

(a) Only (iii) is true

(b) (i) (ii) and (iii) are all type

(c) Only (i) is true

(d) Only (ii) is true

15. Consider a transformer on no load excited at rated voltage. A narrow air-gap is now cut across the yoke of its core. What will happen to the maximum core flux density and magnetizing current? Assume winding resistance and leakage to be negligible.

Core flux density Magnetizing current

(a) Will remain same will increase

(b) Will decrease will decrease

(c) Will decrease will increase

(d) Will increase will remain same.

16. For large power transformers, it is preferable to use

(a) Radiator tanks (b) Corrugated tanks

(c) Tubed tanks (d) Tanks with separate coolers.

17. In general, for all sizes of distribution transformer, it is preferable to use

(a) Tubed tanks (b) Corrugated tanks

(c) Plain sheet steel tanks (d) Radiator tanks.

18. For large power transformers best utilization of available core space can be made by using

(a) Square core section (b) Stepped core section

(c) Rectangular core section (d) None of the above.

19. The core in a large power transformer is built of

(a) Mild steel (b) Ferrite

(c) Silicon steel (d) Cast iron.

20. In the ideal transformer (IT) shown in Fig. below the resistance seen from the HV side is

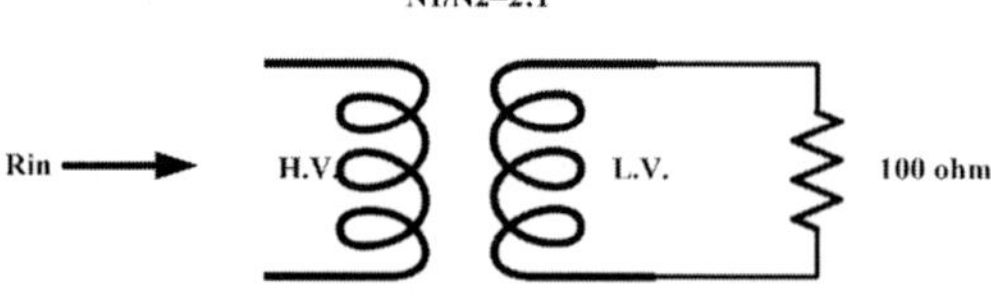


(a) 400 Ω (b) 50 Ω

(c) 25 Ω (d) 100 Ω

21. For fixed primary voltage how does the secondary terminal voltage of a transformer change if the pf of capacitive load rises from 0.7 to 0.9 with load current remaining fixed.

    (a) Voltage

    (b) No change in voltage

    (c) Voltage decreases

    (d) Voltage decreases, goes through a minimum and then rises.

22. The magnetic properties of silicon steel crystal are best

    (a) Along the surface diagonals (b) Along the cube edge

    (c) Along the cube diagonals (d) None of the above

23. A Δ/Y transformer has a phase-to-phase voltage transformation ratio of "a" (delta phase):1 (star phase). The line-to-line voltage ratio Y/Δ is given by

    (a) $\frac{\sqrt{3}}{a}$ (b) $\frac{a}{\sqrt{3}}$

    (c) $\frac{a\sqrt{3}}{1}$ (d) a

24. The magnetic properties of C.R.G.O.S are excellent

    (a) At right angles to the direction of rolling

    (b) Along the direction of rolling

    (c) Against the direction of rolling

    (d) None of the above.

25. Silicon steel is preferred for transformer core because

    (a) It decreases the tensile strength

    (b) It reduces resistivity of the core

    (c) It reduces both eddy current loss and hysteresis loss

    (d) It decreases the permeability of the core.

26. For C.R.G.O.S. mitred overlap is preferable for the core-yoke joints as

    (a) It reduces magnetizing current and also the core-losses

    (b) It improves mechanical strength

    (c) It makes better use of core space

    (d) It reduces magnetostriction.

27. A transformer excited from a sinusoidal voltage source will draw a no-load current:

    (a) Which comprises the fundamental frequency and the depressing third harmonic

    (b) Which is sinusoidal and of the same frequency as the voltage source

    (c) Which comprises the fundamental frequency (same as that of the voltage source) and the peaking third harmonic

    (d) Which is sinusoidal of frequency three times that of the voltage source.

28. When a two-winding transformer is connected as an auto transformer its efficiency (full-load)

    (a) Increases (b) Remains the same

    (c) Decreases (d) Rises to 100%,

29. A 2500/250 V, 25 kVA transformer is connected as an auto transformer to give 2300/2750 V. Its rating could be

    (a) 275 kVA; 250 kVA is transformed inductively

    (b) 250 kVA; 225 kVA is transferred inductively

    (c) 275 kVA; 250 kVA is transformed conductively

    (d) 250 kVA; 225 kVA is transformed conductively.

30. The voltage applied to a transformer primary is increased so as to keep V/f fixed. How will the core loss and magnetizing current change?

Core loss Magnetizing current

    (a) Will increase will remain name

    (b) Will decrease will increase

    (c) Will remain same will decrease

    (d) will remain same will remain same.

31. Spiral winding is suitable only for

    (a) Windings rated for low voltage

    (b) Windings rated for high voltage

    (c) Windings carrying very high current

    (d) Windings carrying very low current.

32. In a transformer the phase angle between primary and secondary terminal voltage is

    (a) $90^0$; primary voltage lagging the secondary voltage

    (b) a few degrees: primary voltage leading the secondary

    (c) a few degrees; primary voltage lagging the secondary voltage

    (d) $90^0$; primary voltage leading the secondary voltage

33. The direction of the central phase winding of a three-phase shell type transformer is reversed with respect to the outer phases

    (a) To reduce short circuit forces

    (b) To minimize eddy current loss

    (c) To save considerable amount of core material

    (d) To reduce leakage fluxes.

34. Helical winding is suitable for

    (a) Low voltage winding of large transformers

    (b) Low voltage winding of small transformers

    (c) High voltage winding of small transformers

    (d) High voltage winding of large transformers

35. In an ideal transformer, the impedance transforms from one side to the other

    (a) In direct ratio of turns

    (b) In inverse ratio of turns

    (e) In direct ratio of square root of turns

    (d) In direct square ratio of turns.

36. Distribution transformers have core losses

    (a) Negligible compared to full-load copper loss

    (b) Less than full-load copper loss

    (c) Equal to full-load copper loss

    (d) More than full-load copper loss.

37. Two transformers connected in parallel share load in the ratio of their kVA ratings only if their PU impedance (on their own kVA's) are
    (a) In inverse ratio of their ratings
    (b) Purely reactive
    (c) In direct ratio ratings
    (d) Equal.
38. Two transformers connected in parallel share loads in the ratio of their kVA ratings provided their impedances are
    (a) In inverse ratio of their ratings
    (b) Purely reactive
    (c) In direct ratio ratings
    (d) Equal
39. Five limb core construction has an advantage over the three-limb core construction that
    (a) The magnetic reluctance of the three phases can be balanced
    (b) The eddy current loss is less
    (c) The permeability is higher
    (d) The hysteresis loss is less.
40. Cross-over winding is suitable for
    (a) High voltage winding of large transformers
    (b) Low voltage winding of large transformers
    (c) Low voltage winding of small transformers
    (d) High voltage winding of small transformers.
41. For distribution transformers of rating less than 50 KVA it is better to use
    (a) Corrugated tanks only
    (b) Either plain sheet steel tanks or corrugated tanks
    (e) Tubed tanks only
    (d) Plain steel tanks only.
42. The use of higher flux density in the transformer design
    (a) Decreases the weight of copper/kVA but decreases that of iron
    (b) Decreases the total weight/kVA
    (c) Increases the total weight/kVA
    (d) Decreases the weight of iron/kVA but increases that of copper.

43. The high frequency hum in the transformer is mainly due to

(a) Magnetostriction

(b) Laminations being not sufficiently tight

(c) Tank walls

(d) Oil of the transformer.

44. Under balance load conditions, the main transformer rating in the Scott connection is greater than that of the teaser transformer by:

(a) 57.7%
(b) 1 5%

(c) 5%
(d) 85%.

45. Tapping's are normally provided on the high voltage of a transformer:

(a) Only because it is easily accessible physically

(b) Only because it has to handle low currents

(c) Because it has larger number of turns, has to handle low currents and is also easily accessible physically

(d) Only because it has large number of turns which allows smoother variation of voltage

46. Minor insulation of a transformer is the insulation

(a) between the turns of the windings only

(b) between the low-voltage and the high-voltage winding

(c) between the turns of the windings and also between the layers of the windings

(d) between the layers of the windings only.

47. Transformer oil is used as

(a) inert medium
(b) coolant only

(c) both insulant and coolant
(d) insulant only.

48. Buchholz relay is placed

(a) in between 1.v. winding and the bushing

(b) in between the conservator and the breather

(c) in between the tank and the conservator

(d) in between the h.v. winding and the bushing.

49. A transformer may have two or more ratings depending upon

(a) the type of insulation used
(b) the type of winding used
(c) the type of core used
(d) the type of cooling used.

50. A single-phase transformer is feeding a lagging load. What will be the effect on the voltage regulation of the transformer if:

(i) an inductor is connected in parallel to the load
(ii) a capacitor is connected in parallel to the load

(a) Regulation will increase in both the cases
(b) Regulation will increase in case (i) but decrease in case (ii)
(c) Regulation will decrease in case (i) but increase in case (ii)
(d) Regulation will decrease in both the cases.

51. A 400/200 V transformer has a PU impedance of 0.05. The HV aide voltage required to circulate full load current during short-circuit test is

(a) 20 V
(b) 5 V
(c) 40 V
(d) 10 V

52. Buchholz relay is

(a) a gas actuated device
(b) a voltage sensitive device
(c) a current sensitive device
(d) a frequency sensitive device.

53. $P_i$ =core loss, $P_c$ =copper loss. A transformer has maximum efficiency when:

(a) $P_i/P_c = 2$
(b) $P_i/P_c = 1.5$
(c) $P_i/P_c = 0.5$
(d) $P_i/P_c = 1.0$.

54. A transformer on no-load is switched on to a source of voltage. It will draw a current.

(a) which is several times the steady-state magnetizing current depending upon the initial state of the residual flux in the transformer core
(b) which is twice the steady-state magnetizing current providing the core has no residual flux
(c) which is several times the steady-state magnetizing current, independent of the initial state of residual flux in the transformer core
(d) which is the same as the steady-state magnetizing current.

55. A transformer has maximum efficiency at full-load. Compare its all-day efficiency when operated

(i) full-load all the time

(ii) full-load half time and half load half time

(iii) full-load half time and no-load half time

Mark the correct answer below:

(a) Efficiency will be highest in case (i) and lowest in case (ii)

(b) Efficiency will be lowest in case (i) and highest in case (iii)

(c) Efficiency will be highest in case (i) and lowest in case (iii).

56. Sludging of transformer oil means.

(a) formation of semi-solid hydro-carbon due to heat and oxidation

(b) continuous expansion and contraction due to heating and cooling

(c) evaporation of transformer oil due to heating

(d) decomposition of transformer oil under the influence of power arcs.

57. A transformer operates most efficiently at 3/4th full-load. Its iron loss ($P_i$) and full load copper loss ($P_c$) are related as

(a) $P_i/P_c = 4/3$ (b) $P_i/P_c = 16/9$

(c) $P_i/P_c = 9/16$ (d) $P_i/P_c = 3/4$

58. A 2/1 ratio, two-winding transformer is connected as an auto transformer. Its kVA rating as an auto transformer compared to a two-winding transformer is

(a) 3 times (b) same

(c) 1.5 times (d) 2 times

59. R = equivalent resistance, X equivalent reactance, $P_i$= core loss. The load current for maximum efficiency operation of a transformer is given by

(a) $\sqrt{\frac{P_i}{R}}$ (b) $\frac{P_i}{X}$

(c) $\frac{P_i}{R}$ (d) $\sqrt{\frac{P_i}{X}}$

60. Which of the following tests must be performed on a transformer to determine its leakage reactance?

(a) OC test only (b) both OC and SC tests

(c) SC test only (d) test by an impedance bridge

61. Why is the yoke cross-section made larger than core cross-section in a transformer?

(a) to reduce core loss

(b) to provide a better mechanical fit between yoke and core

(c) to reduce magnetizing current

(d) to increase mechanical strength

62. Why are transformer stampings annealed before being used for the core building?

(a) to reduce hysteresis loss due to burring of edges

(b) to give mechanical strength to the core

(c) to increase core permeability

(d) to reduce eddy-current loss due to burring of edges.

63. A transformer has negligible resistance and a per unit reactance of 0.1. Its voltage regulation on full-load with a leading pf angle of $30^0$ leading is:

(a) -5% (b) 10%

(c) -10% (d) 5%

64. R = equivalent resistance, X = equivalent reactance and cosφ = pf leading. The voltage rises of a transformer upon throwing off full-load current:

(a) $I_{fl} R\sin j + I_{fl} X\cos j$ (b) $I_{fl} R\sin j + I_{fl} X\cos j$

(c) $I_{fl} R\sin j + I_{fl} X\cos j$ (d) $I_{fl} R\sin j + I_{fl} X\cos j$

65. The applied voltage of a certain transformer is increased by 50% while the frequency is reduced to 50%. The maximum core flux density will become

(a) 1.5 times (b) 0.5 times

(c) 3 times (d) will remain the same

66. On the two sides of a star/delta transformer

(a) the voltage and currents both differ in phase by $30^0$

(b) the voltage and currents are both in phase

(c) the currents differ in phase by $30^0$ but voltages are in phase

(d) the voltage and currents are both in phase.

67. A 200/100 V, 50 Hz transformer is to be excited at 40 Hz from the 100 V side. For the exciting current to remain the same, the applied voltage should be

(a) 80 V
(b) 150 V
(c) 100 V
(d) 125 V

68. A 6600 V, 50 Hz transformer operates at a flux density of 1.5 T. Each linear dimension of the secondary turns is halved. If the transformer is now operated at 13200 V, 50 Hz, what will be the core flux density?

(a) 3 T
(b) 6 T
(c) 1.5 T
(d) 4.5 T

69. During SC test at full-load current the power input to a transformer comprises predominantly

(a) eddy current loss
(b) core loss
(c) copper loss
(d) both core and copper loss.

70. Conservator is used

(a) for better cooling of the transformer
(b) to act as an oil storage
(c) to take up the expansion of oil due to temperature rise
(d) none of the above.

71. Sumpner's test on two identical transformers yields information about

(a) it yields no information on losses
(b) core loss only
(c) both core loss and full-load copper loss
(d) full-load copper loss only.

72. The voltage regulation of a transformer at full-load .85 pf lagging is 5%. Its voltage regulation at full-load .85 pf leading.

(a) will reduce and may even become negative
(b) will remain the same
(c) will be positive
(d) will be negative.

73. The oil used in the transformer should be free from moisture because moisture will
    - (a) cause its lubricating property to deteriorate
    - (b) reduce its density
    - (c) reduce its dielectric strength
    - (d) cause the transformer core to rust.

74. The power transformer is a
    - (a) Constant current device
    - (b) Constant voltage device
    - (c) Constant main flux device
    - (d) Constant power device.

75. Power input to a transformer on no load at rated voltage comprises predominantly
    - (a) core loss
    - (b) eddy current loss
    - (c) copper loss
    - (d) hysteresis loss.

76. What is the arrangement of windings in a core type single-phase transformer?
    - (a) half HV inside and half HV outside each core limb
    - (b) sandwiched LV and HV discs on each core limb
    - (c) half LV inside and half HV outside on each core-limb
    - (d) LV on one core limb and HV on the other.

77. Why are core bolts insulated in a transformer?
    - (a) to prevent bolt corrosion
    - (b) to prevent increase of eddy current loss
    - (c) to prevent noise caused by bolt vibration
    - (d) to prevent increase of hysteresis loss.

78. The color of fresh dielectric oil for a transformer is
    - (a) pale yellow
    - (b) dark brown
    - (c) grey
    - (d) colorless.

## ANSWERS

| 1. (b) | 2. (b) | 3. (a) | 4. (c) | 5. (c) |
|---|---|---|---|---|
| 6. (a) | 7. (c) | 8. (b) | 9. (6) | 10. (a) |
| 11. (c) | 12. (c) | 13. (c) | 14. (a) | 15. (a) |
| 16. (d) | 17. (a) | 18. (b) | 19. (c) | 20. (a) |
| 21. (c) | 22, (b) | 23. (a) | 24. (b) | 25. (c) |
| 26. (a) | 27. (c) | 28. (a) | 29. (c) | 30. (a) |
| 31. (c) | 32. (b) | 33. (c) | 34. (a) | 35. (d) |
| 36. (b) | 37. (c) | 38. (a) | 39. (a) | 40. (d) |
| 41. (b) | 42. (b) | 43. (a) | 44. (b) | 45. (c) |
| 46. (e) | 47. (c) | 48. (c) | 49. (d) | 50. (b) |
| 51. (a) | 52. (a) | 53. (d) | 54. (a) | 55. (c) |
| 56. (a) | 57. (c) | 58. (a) | 59. (a) | 60. (c) |
| 61. (c) | 62. (d) | 63. (a) | 64. (b) | 65. (c) |
| 66, (a) | 67. (a) | 68. (c) | 69. (c) | 70. (c) |
| 71. (c) | 72. (a) | 73. (c) | 74. (c) | 75. (a) |
| 76. (c) | 77. (b) | 78. (a) | | |

## II. D.C. MACHINES

1. Interpoles in a de motor must be

(a) Series excited and should have the same polarity as that of the next main pole in the direction of rotation of the armature.

(b) Series excited and should have polarity opposite to that of the next main pole in the direction of rotation of armature.

(c) Shunt excited and should have the same polarity as the next main pole in the direction of rotation of the armature.

(d) Shunt excited and should have polarity opposite to that of the next main pole in the direction of rotation of the armature.

2. Why are pole tips in a de machine chamfered?

(a) To increase induced emf per coil.

(b) To improve commutation characteristics.

(c) To reduce armature reaction effect.

(d) To achieve nearly sinusoidal air-gap flux density distribution. .

3. Let θ be the angle between the sinusoidally distributed stator and rotor mmf's then

(a) In d.c. synchronous and induction machines θ is fixed at 90°.

(b) In d.c. machines θ is fixed at 90° while θ in synchronous and induction machines can have different values.

(c) In d.c. synchronous and induction machines θ can be different from 90°.

(d) In synchronous and induction machines θ is fixed at 90°, while θ in d.c. machines can have different value.

4. A homopolar generator usually has

(a) Low voltage and high current. (b) High voltage and low current.

(c) High voltage and high current. (d) Low voltage and low current.

5. A conductor of length l metre is perpendicular to a magnetic flux system of density B Wb/m² and carries a current of I amp. It experiences a force of

(a) 0.1 BIl dynes in the direction perpendicular to both B and l.

(b) BIlNW in the direction of B.

(c) BIlNW in the direction perpendicular to both B and l.

(d) BIlNW in the direction of l.

6. The brushes of a d.c. machine should be placed

(a) on the commutator in the polar axis.

(b) on the commutator in the interpolar axis.

(c) on the armature in the interpolar axis (midway between poles).

(d) on the armature in the polar axis.

7. DC machines poles are constructed of thick laminations

(a) to reduce iron loss in pole body and for ease of construction.

(b) to reduce pulsation loss in pole shoes and for ease of construction.

(c) for ease of construction.

(d) to reduce iron loss in pole body and pole shoes.

8. The armature reactions AT in a dc machine

(a) make an angle of 90° with the main pole axis.

(b) are in the same direction as the main poles.

(c) are in direct opposition to the main poles.

(d) make an angle with the main pole axis which is load dependent.

9. In a DC machine

   (a) EMF's in armature conductors as well as at terminals are unidirectional.

   (b) Current in armature conductors and at the terminals is alternating while e.m.f. is unidirectional.

   (c) The current and e.m.f. in armature conductors are alternating while those at the terminals are unidirectional.

   (d) The current and the e.m.f. in armature conductors as well as at terminals are unidirectional.

10. The sinusoidally distributed stator and rotor m.m.f. in the uniform airgap of a machine have fixed amplitudes. The torque will be maximum when the angle between them is

    (a) 45° (b) 90°

    (c) 30° (d) 0°.

11. A straight horizontal conductor falls vertically under the action of gravity cutting perpendicularly horizontal magnetic lines of force. The conductor is short-circuited at its ends with end connections not cutting any flux line. It has

    (a) gradually increasing acceleration.

    (b) Gradually decreasing acceleration.

    (c) Constant acceleration.

    (d) No acceleration at all

12. Open slot is used in de machine armature because

    (a) it reduces the coil reactance e.m.f. and hence aids in commutation.

    (b) of the ease with which the winding can be placed inside the slots.

    (c) it reduces the armature voltage drop.

    (d) it increases the induced e.m.f. per coil.

13. The motor best suited for hoist, crane and traction type load is the

    (a) d.c. series motor. (b) synchronous motor.

    (c) induction motor (d) de shunt motor.

14. M.K.S. units of magnetic flux, flux density and intensity are respectively

    (a) Maxwell, Gauss, Amp turns.

    (b) Weber, Weber/metre, Amp-turns.

(c) Weber, Weber/metre, Amp/metre.

(d) Lines, Lines/metre, Oersted.

15. A short-circuited rectangular coil falls under gravity with the coil remaining in a vertical plane and cutting perpendicularly horizontal magnetic lines of force. Its acceleration

(a) Increases (b) Decreases

(e) Remains constant (d) is zero.

16. The induced e.m.f. in a conductor of length l moving with velocity v in magnetic flux of density B, while B, I and v are mutually perpendicular, is Blv volts provided

(a) B is in lines/metre, I is in metre and v is in metre/sec.

(b) B is in Weber, I is in metre and v is in metre/second.

(c) B is in Weber/metre$^2$, I is in metre and u is in metre per second.

(d) B is in gauss, I is in centimeter and v is in centimeter/second.

17. Which one of the following statements is correct?

(a) The direction of current in a generator is same as that of its induced e.m.f. while that in a motor is opposite to its induced e.m.f.

(b) The direction of current in a generator is opposite to its induced e.m.f. while that is a motor is same as that of its induced e.m.f.

(c) The direction of currents in both generators and motors is same as that of their induced e.m.f

(d) The direction of currents in both generators and motors is opposite to their induced e.m.fs..

18. What losses occur in the teeth of a d.c. machine armature?

(a) Both hysteresis and eddy current loss.

(b) Eddy current loss only.

(c) Hysteresis loss only.

(d) No losses.

19. Upon loading a d.c. machine flux/pole:

Machine operating in unsaturating region Machine operating in saturated region

(a) remains constant decreases

(b) first decreases and then increases remains constant

(c) decreases remains constant

(d) increases first decreases and then increases.

20. Inter-poles help commutation in a d.c. machine by

(a) aiding the main poles.

(b) cancelling the armature reaction m.m.f.

(c) by causing dynamically induced e.m.f. in the coils undergoing commutation.

(d) by causing statically induced e.m.f. in the coils undergoing commutation.

21. In a d.e. machine which is operating in the saturated region, the armature reaction effect is:

(a) demagnetizing only.

(b) cross-magnetizing as well as demagnetizing.

(c) magnetizing only.

(d) cross-magnetizing only.

22. In a duplex wave winding with equalizers

(a) number of pole-pairs is even and number of slots is odd

(b) both number of pole-pairs and numbers of slots are odd.

(c) both number of pole-pairs and number of slots are even.

(d) number of pole-pairs is odd and number of slots is even.

23. Mark the completely correct statement below in respect of d.c. series motor. Assume negligible armature circuit resistance, the magnetic circuit is linear

(a) The speed varies inversely as the square of the load torque, at a given torque load speed, is directly proportional to the number of field turns.

(b) The speed varies inversely as the square root of the load torque, at a given torque load speed, is inversely proportional to the number of field turns.

(c) The speed varies inversely as the load torque, at a given torque load speed, is inversely proportional to the number of field turns.

(d) The speed varies inversely as the square root of the load torque, at a given torque load speed, is directly proportional to the number of field turns.

24. If a DC series motor is connected to an a.c. supply, it will exert
   (a) high torque (b) zero torque
   (c) unidirectional torque (d) pulsating torque.
25. The number of reentrancy of a triplex wave winding is 3 only if
   (a) The number of slots is divisible by 3.
   (b) The number of coils is divisible by 3.
   (c) The commutator pitch is divisible by 3.1
   (d) Both the number of coils and commutator pitch are divisible by 3.
26. In a duplex wave winding with equalizers
   (a) both number of pole-pairs and number of slots are even.
   (b) number of pole-pairs is odd and number of slots is even.
   (c) number of pole-pairs is even and number of slots is odd.
   (d) both number of pole-pairs and number of slots are odd.
27. The teeth in the armature of a d.c. machine are sometimes skewed
   (a) to reduce eddy current loss.
   (b) to reduce vibrations.
   (c) to reduce both hysteresis and eddy current loss.
   (d) to reduce hysteresis loss.
28. The armature iron is laminated
   (a) to reduce cost.
   (b) to reduce hysteresis loss.
   (c) to reduce eddy current loss.
   (d) to reduce both hysteresis and eddy current loss.
29. The number of reentrancy of a duplex lap winding is 2 only if
   (a) The number of slots is odd. (b) The number of coils is even.
   (c) The commutator pitch is 2. (d) The commutator pitch is odd.
30. Why is the pole shoe in a d.c. machine larger than its pole body?
   (a) It helps to make the flux density wave nearly sinusoidal.
   (b) It gives sinusoidal flux density.

(c) It reduces iron loss in the pole shoes and gives a more nearly rectangular flux density wave.

(d) It provides a support for field winding.

31. While starting a d.c. shunt motor

(a) reduced armature voltage but full field voltage should be applied and all regulator resistance should be cut out in the field circuit.

(b) rated armature and rated field voltage should be applied and full regulator resistance should be included in the field circuit.

(c) reduced armature voltage and reduced field voltage should be applied and full regulator resistance should be included in the field circuit.

(d) rated armature and rated field voltage should be applied and full regulator resistance should be cut out in the field circuit.

32. In a d.c. motor the torque (Nm) developed is

(a) $W_m/E_aI_a$, in the direction of $W_m$

(b) $E_aI_a / W_m$ in the direction of $W_m$

(c) $W_m / E_aI_a$ opposite to the direction of $W_m$ (speed in mech, rad/5).

(d) $E_aI_a / W_m$ opposite to the direction of $W_m$

33. In a drum type d.c. armature winding the back pitch and front pitch in terms of coil sides must be

(a) even and odd respectively

(b) odd and even respectively

(c) both even

(d) both odd.

34. Mark the completely correct statement below in respect to a d.c. shunt motor

(a) No-load speed is directly proportional to flux/pole and inversely proportional to armature voltage, the speed rises linearly with load torque.

(b) No-load speed is inversely proportional to flux/pole and directly proportional to armature voltage, the speed drops off linearly with load torque.

(c) No-load speed is inversely proportional to flux/pole and directly proportional to armature voltage, the speed rises linearly with load torque.

(d) No-load speed is directly proportional to flux/pole and inversely proportional to armature voltage, the speed drops off linearly with load torque.

35. What would be observed if a d.c. shunt motor is shorted with an open-circuited field?

    (a) The motor picks up fast and acquires full speed while drawing small current.

    (b) The motor picks up fast and acquires full speed while drawing large current,

    (c) The motor does not pick up speed but draws a large current

    (d) The motor does not pick up speed but draws a small current.

36. A duplex lap winding with equalizers and even number of pairs of poles is

    (a) singly re-entrant and number of coils per pair of poles is an odd integer.

    (b) singly re-entrant and number of coils per pair of poles is an even integer.

    (c) doubly re-entrant and number of coils per pair of poles is an even integer.

    (d) doubly re-entrant and number of coils per pair of poles is an odd integer.

37. A 6-pole d.c. armature has simplex lap-connected 720 conductors, 3 turns per coil and 4 coil-sides per slot. Determine the number of slots in the armature and state whether equalizers can be employed

    (a) 120 slots, not possible. (b) 60 slots, not possible.

    (c) 30 slots, possible (d) 60 slots, possible.

38. The commutator pitches of simplex and duplex lap windings are respectively

    (a) 1 and 1. (b) 2 and 2.

    (c) 1 and 2. (d) 2 and 4.

39. A differentially compounded motor under high over-load conditions will behave like a/an.

    (a) a.c. synchronous motor. (b) Shunt motor.

    (c) Series motor. (d) Cumulative compound motor.

40. Why is the armature of a d.c. machine made of silicon steel stampings?

    (a) To reduce eddy current loss.

    (b) To reduce hysteresis loss.

    (c) For the ease with which the slots can be created.

    (d) To achieve high permeability.

41. The following condition is to be satisfied to avoid split coils in drum type d.c. armature windings

(a) ——— = an integer (b) $\frac{Y_b - 1}{u}$ an odd integer

(c) $\frac{Y_b - 1}{u}$ = an integer (d) $\frac{Y_b - 1}{u}$ an even integer.

42. Due to magnetic saturation, the flux per pole in a d.c. machine without brush shift

(a) Decreases in the generators and increases in the motors with load.

(b) Increases in the generators and decreases in the motors with load.

(c) Decreases in both the generators and motors with load.

(d) Increases in both the generators and the motors with load.

43. A d.c. machine is provided with both interpole winding (IPW) and compensating winding (CPW) with respect to the armature

(a) both IPW and CPW are in series.

(b) both IPW and CPW are in parallel.

(c) IPW is in parallel and CPW is in series.

(d) IPW is in series and CPW is in parallel.

44. Consider the following statements

The purpose of using interpoles in machines is to counteract

(i) The demagnetizing effect of armature m.m.f in the commutating zone

(ii) The cross-magnetizing effect of armature m.m.f in the commutating zone.

(iii) The reactance voltage.

State which of the following is correct:

(a) Only (ii) and (iii) (b) Only (i) and (iii)

(c) Only (ii) (d) Only (iii).

45. For a d.c. generator if the brushes are given a small amount of forward effect of armature reaction is

(a) Totally cross-magnetizing.

(b) Totally demagnetizing.

(c) Partly demagnetizing and partly cross-magnetizing

(d) Totally magnetizing.

46. The brushes of a d.c. machine are

    (a) Physically placed in the polar axes and electrically connected to the coils in the polar axes.

    (b) Physically placed in the polar axes and electrically connected to the coils in the interpolar axes.

    (c) Physically placed in the interpolar axes and electrically connected to the coils in the interpolar axes.

    (d) Physically placed in the interpolar axes and electrically connected to the coils in the polar axes.

47. Slot wedges in a d.c. machine are made of

    (a) Silicon steel. (b) Fibre

    (c) Mild steel. (d) Cast iron.

48. A 6-pole lap wound d.e. armature with 720 conductors draws 50 A from the mains. If armature reaction per pole is

    (a) 500 AT peak, sinusoidal in wave shape,

    (b) 1000 AT peak, sinusoidal in wave shape

    (c) 500 AT peak, triangular in wave shape.

    (d) 1000 AT peak, triangular in wave shap

49. The armature reaction m.m.f. wave in a d.e. machine

    (a) Moves relative to the brushes at armature speed.

    (b) Is stationary relative to the brushes.

    (c) Moves relative to the brushes at synchronous speed.

50. Field control of a d.e. shunt motor gives

    (a) Constant kW drive (b) Constant torque drive

    (c) Constant speed drive (d) Variable load speed drive.

51. The commutator of a d.c. machine acts as a

    (a) Controlled full-wave rectifier. (b) Half-wave rectifier.

    (c) Full wave rectifier. (d) Controlled half-wave rectifier.

52. In a d.c. machine the armature m.m.f. is always directed along the

    (a) Brush axis (b) Polar axis

    (c) Interpolar axis (d) None of the above.

53. Compared to an uncompensated d.c. machine, the interpole AT required in a compensated d.c. machine is

    (a) Large for both the generators and the motors.

    (b) Large for the generators and smaller for the motors

    (c) Smaller for the generators and larger for the motors.

    (d) Smaller for both the generators and the motors.

54. Consider the following statements

    (i) Stationary with respect to the field poles.

    (ii) Rotating with respect to the field poles.

    (iii) Rotating with respect to the armature.

State which of the following is correct:

(a) Only (iii) (b) Only (i) and (iii)

(c) Only (i) (d) Only (ii) and (iii)

55. Consider the following statements

    (i) Armature reaction m.m.f. under the pole-faces.

    (ii) Armature reaction m.m.f. in the interpolar zone.

    (iii) Flashover between positive and negative brushes,

State which of the following is correct.

(a) Only (i) and (ii) (b) Only (ii) and (iii)

(c) Only (i) and (iii) (d) Only (iii).

56. The waveform of the armature m.m.f. in a d.c. machine is

    (a) rectangular (b) sinusoidal

    (c) square (d) triangular.

57. If the armature current of a d.c. motor is increased keeping the field flux constant, then the developed torque

    (a) Remains constant.

    (b) Increases proportionally.

    (c) Decreases in inverse proportion.

    (d) Increases proportional to the square of the current.

58. In d.c. machines, the polarity of the interpole is
    (a) Same as that of the main pole behind for both the generators and the motors.
    (b) Same as that of the main pole behind for the generators and that of the main pole ahead for the motors.
    (c) Same as that of the main pole ahead for the generators and that of the main pole behind for the motor.
    (d) Same as that of the main pole ahead for both the generators and the motors
59. Due to magnetic saturation, the flux per pole in a d.c. machine without brush shift
    (a) Increases in both the generators and the motors with load.
    (b) Decreases in the generators and increases in the motors with load.
    (c) Increases in the generators and decreases in the motors with load.
    (d) Decreases in both the generators and the motors with load.
60. Compared to the air-gap under the field-poles, the interpole air-gap is made
    (a) Smaller for generators and larger for the motors.
    (b) Smaller for both the generators and the motors.
    (c) Larger for both the generators and the motors.
    (d) Larger for the generators and smaller for the motors.
61. In a d.c. generator, e.m.f. due to residual flux is fed back to the shunt winding positively. Why does the no-load voltage build up to a finite steady value?
    (a) Because of eddy-current loss in pole shoes.
    (b) Because of field winding resistance.
    (c) Because of magnetic saturation.
    (d) Because of field winding inductance.
62. The armature reaction m.m.f. in a d.c. machine is
    (a) Triangular in shape. (b) Sinusoidal in shape.
    (c) Trapezoidal in shape. (d) Rectangular in shape.
63. In a d.c. machine without interpoles to get improved commutation, the brush shift angle must be
    (a) Kept constant (b) 0°
    (c) Varied with change in load (d) None of the above.

64. Consider the following statements:

Compensating windings are used in a d.c. motor which are intended to operate

(i) With rapidly changing loads of wide range.

(ii) At constant speed over wide range of load.

(iii) Over wide range of speed by field excitation control.

State which of the following is correct:

(a) Only (iii)
(b) Only (i)
(c) Only (i) and (iii)
(d) Only (ii) and (iii).

65. A d.c. series motor has linear magnetization and negligible armature resistance. The motor speed is

(a) Inversely proportional to T.
(b) Inversely proportional to $T^{0.5}$
(c) Directly proportional to T.
(d) Directly proportional to $T^{0.5}$

66. In Hopkinson's test on two identical shunt motors the power input to the armature circuit comprises

(a) Armature copper loss + windage and friction loss

(b) Copper loss + no-load iron loss + stray-load iron + windage and friction loss.

(c) Armature copper loss + no-load iron loss + windage and friction loss.

(d) Armature copper loss + no-load iron loss + stray-load iron loss.

67. A d.c. machine has maximum efficiency when

(a) Constant losses equal to variable losses

(b) Windage and friction losses equal to copper loss.

(e) Constant losses equal to losses proportional to the square of the current.

(d) Iron losses equal to copper losses.

68. Under-commutation gives rise to

(a) Sparking at the trailing edge of the brush.

(b) Sparking at the middle of the brush.

(c) No sparking at all.

(d) Sparking at the leading edge of the brush.

69. In a d.c. machine without any brush shift, the shift of the magnetic neutral axes due to armature reaction is

    (a) Against the directions of rotation for both the generator and the motor.

    (b) Against the direction of rotation for the generator and in the direction of rotation for the motor.

    (c) In the direction of rotation for both the generator and the motor.

    (d) In the direction of rotation for the generator and against the direction of rotation for the motor.

70. The e.m.f. induced in the armature of a d.c. machine is

    (a) Directly proportional to both the flux and the speed.

    (b) Inversely proportional to both the flux and the speed.

    (c) Directly proportional to the flux and inversely proportional to the speed.

    (d) None of the above.

71. The direction of induced e.m.f. in an armature coil of a d.c. machine is

    (a) Opposite to that of the current for both the generator and the motor.

    (b) The same as that of the current for both the generator and the motor.

    (c) The same as that of the current for the generator and opposite to that current the motors.

    (d) None of the above.

72. The parts of the armature electric circuit which take active part in e.m.f. generation are

    (a) The commutator segments.

    (b) The overhangs.

    (c) The coil-sides inside the slots.

    (d) Both the coil sides inside the slots and the overhangs.

73. In a d.c. motor if the brushes are given a backward shift, then

    (a) Commutation is improved and speed increases.

    (b) Commutation is unaffected and speed increases.

    (c) Commutation is improved and speed decreases.

    (d) Commutation is worsened and speed decreases.

74. A d.e. shunt motor while running on no-load draws power mainly for

(a) No-load iron loss stray-load iron loss windage and friction loss.

(b) No-load iron loss + windage and friction loss.

(c) No-load iron loss only.

(d) Windage and friction loss only.

75. Mark the correct statement below in respect of the field control of a d.c. shunt motor

(a) Weakening the field reduces motor speed but increases its full-load torque capability.

(b) Weakening the field reduces motor speed but does not affect its full-load torque capability.

(c) Weakening the field increases motor speed but reduces its full-load torque capability.

(d) Weakening the field increases motor speed but does not affect its full-load torque capability.

76. The critical resistance in a d.c. shunt generator is

(a) The value of field circuit resistance above which the generator would fail to excite.

(b) The value of field circuit resistance for which the generator no-load voltage equals the rated voltage.

(c) The resistance of the field circuit.

(d) The value of field circuit resistance below which the generator would fail to excite.

77. A d.c. shunt generator has been developing rated voltage at rated speed. Match the statements in lists I and II and select correct answer using the code below.

| **List I** | **List II** |
|---|---|
| A. The direction of rotation and residual magnetism are reversed. | I. The generator voltage will not build up |
| B. The direction of rotation and connections of the field winding are reversed. | II. The generator voltage will build up with same polarity |
| C. The direction of residual magnetism and field connections are reversed. | III The generator voltage will build up with reversed polarity |
| D. The direction of rotation, residual magnetism and field connections are reversed. | |

Code:

| (a) | A<br>III | B<br>I | C<br>II | D<br>II | (b) | A<br>I | B<br>II | C<br>I | D<br>III |
|---|---|---|---|---|---|---|---|---|---|
| (c) | A<br>I | B<br>III | C<br>I | D<br>II | (d) | A<br>III | B<br>III | C<br>I | D<br>II |

78. A DC generator is run with different modes of field excitation keeping the no-load voltage constant at 250 V every time. The full-load voltages for the various connections are 250, 230, 215 and 180 V. Then the machines have following modes of excitation respectively.

(a) Cumulative compound, separately excited, shunt and differential compound.

(b) Cumulative compound, separately excited, differential compound and hunt.

(c) Separately excited, cumulative compound, shunt and differential compound.

(d) Cumulative compound, differential compound, separately excited and hunt.

79. In a series-parallel field control of a DC series motor with fixed armature current

(a) Series connection gives higher speed.

(b) Both series and parallel connection give the same speed.

(c) parallel connection gives higher speed.

(d) Such connections are not used in practice.

80. The process of current commutation in a DC machine is opposed by the

(a) Reactance e.m.f.

(b) Coil resistance.

(c) Brush resistance

(d) Emf induced in the commutating coil because of the interpole flux.

81. In an unsaturated DC machine armature reaction effect is

(a) Magnetizing

(b) Demagnetizing

(c) Cross-magnetizing.

(d) Kind of effect depends upon whether the machine is motoring or generating.

82. For a given torque, reducing the diverter resistance of a DC series motor

(a) Decreases its speed, demanding less armature current.

(b) Decreases its speed but armature current remains the same.

(c) Increases its speed, demanding more armature current.

(d) Increases its speed but armature current remains the same.

83. Consider the following statements regarding building up of voltage of a shunt generator

(i) There should be residual magnetism in the field system.

(ii) The field winding should be properly connected so that the current in the field winding produces flux in the same direction as that of residual magnetism.

(iii) The resistance of the field winding should be less than the critical resistance corresponding to the speed of the machine.

For build-up voltage.

(a) only (iii) has to be satisfied.

(b) only (i) and (ii) have to be satisfied.

(c) only (i) and (iii) have to be satisfied.

(d) all the above conditions have to be satisfied.

84. Equalizer bus is necessary for the parallel operation of the following types of d.c. generators.

(a) Over compound generators only.

(b) Series generators only.

(c) Series and over compound generators.

(d) Series and any type of compound generators.

85. In a level compound generator the series field ampere turns are

(a) In the same direction as the shunt field ampere turns.

(b) At $90^0$ (elect) to the shunt field ampere turns.

(c) Placed on the interpoles.

(d) in direct opposition to the shunt field ampere turns.

86. The most suitable generator for welding purpose is

(a) Cumulative compound.
(b) Shunt
(c) Differential compound
(d) Separately excited.

87. The following type of DC generators is the most suitable as booster

(a) Series
(b) Shunt
(c) Separately excited
(d) Compound.

88. In a level compound generator the terminal voltage at half-load is

(a) Less than the no-load voltage.

(b) The same as the no-load voltage.

(c) The same as the full-load voltage.

(d) more than the no-load voltage.

89. The voltage on full-load of a DC generator is found to be equal to its voltage on no-load. The generator is

(a) Shunt
(b) Differential compound
(c) Cumulative compound
(d) Separately excited.

90. A DC cumulative compound generator with interpoles was operating satisfactorily supplying steady DC load when the machine was stopped. The machine was re-run with same connections and the same direction of rotation but with the polarity of residual magnetism reversed. Consider the following statements

(i) The machine does not build up.

(ii) The machine builds up with reversed polarity.

(iii) The machine now runs as differential compound generator.

(iv) The interpoles have proper polarity for good commutation.

Select the correct answer from the following code.

(a) Only (ii) and (iii) are true.
(b) Only (i) is true.
(c) Only (ii) and (iv) are true.
(d) Only (ii), (iii) and (iv) are true.

91. Mark the correct statement in Table I in respect of speed-torque characteristics of DC motors

Table I

| | Shunt motor | Series motor | Compound motor (cumulative) |
|---|---|---|---|
| (a) | slightly drooping curve | rectangular hyperbola | heavily drooping curve. |
| (b) | rectangular hyperbola | slightly drooping curve | heavily drooping curve |
| (c) | heavily drooping curve | rectangular hyperbola | slightly drooping curve |
| (d) | heavily drooping curve. | slightly drooping curve. | rectangular hyperbola. |

92. In a shunt motor for given field and armature currents

(a) With unsaturated magnetic circuit the motor can acquire a dangerously high speed.

(b) The state of saturation of the magnetic circuit will have no effect on the speed of the motor.

(c) Speed will be higher if the magnetic circuit is unsaturated than if the magnetic circuit is saturated.

(d) Speed will be higher if the magnetic circuit is saturated than if the magnetic circuit is unsaturated.

93. A DC generator is run with different modes of field excitation keeping no-load voltage constant. The short circuit current is minimum when it is

(a) Sshunt
(b) Separately excited
(c) Differential compound
(d) Cumulative compound.

94. Two coupled DC series motors with constant torque load are changed over from series to parallel connection across a fixed voltage supply. How does the set speed change compare to the original speed?

(a) Speed becomes double.
(b) Speed becomes 1.414 times.
(c) Speed becomes half.
(d) Speed does not change.

95. For build-up of voltage residual magnetism is essential in field systems of all types of DC generators except

(a) Separately excited.
(b) Series and separately excited
(c) Compound and separately excited
(d) Shunt

96. A DC generator running at fixed speed and with fixed shunt field resistance has

(a) Short-circuit current equal to the maximum load current that it can feed

(b) Short-circuit current equal to the full-load current

(c) Short-circuit current more than the maximum load current that it can food

(d) Short-circuit current less than the maximum load current that it can feed

97. On switching a DC motor is found to rotate in the direction opposite to that for which it is designed. The motor is

(a) Shunt
(b) Cumulative compound
(c) Differential compound
(d) Series

98. A cumulative compound motor runs 1000 r.p.m. on no-load. On full-load the flux increases by 20% whereas the full-load drop in the combined resistance of the armature and series field is 4%. What is the full-load speed?

(a) 1240 r.p.m.
(b) 640 rpm.
(c) 800 r.p.m.
(d) 960 r.p.m.

99. A series motor must not be run at low loads because

(a) The speed will be very low.
(b) There will be complete demagnetization of the field system
(c) The speed will be very high.
(d) The current will be very high.

100. Carbon brushes are used in DC machines because

(a) They aid in reversing the commutating coil current only.
(b) They serve the purpose of interpoles.
(c) They aid in reversing the commutating coil current and keep down the wear of the commutator.
(d) They keep down the wear of the commutator only.

101. At full-load the demagnetizing effect of armature reaction reduces the flux by 10%, whereas the full-load armature circuit drop is 5.5%. What is the full-load speed expressed as percentage of no-load speed of the motor?

(a) 110%
(b) 90%
(c) 105%
(d) 94.5%

102. A DC series motor has the following data for its magnetizing characteristic in terms of percentage of full-load values. 80

| Current% | 60 | 80 | 100 |
|---|---|---|---|
| Flux% | 83.33 | 93 | 100 |

The series field has the resistance of .1 ohm. The full-load speed is 1000 rpm. What is the diverter resistance to raise the full-load speed to 1200 r.p.m. full load torque remaining unaltered?

(a) 0.2 ohm
(b) 0.15 ohm
(c) 0.1 ohm
(d) 0.05 ohm.

103. In the block diagram of a separately excited d.c. motor, the armature induced e.m.f. appears as

(a) Disturbance input
(b) Positive feedback
(c) Negative feedback
(d) Output.

104. A DC over-compound generator was operating satisfactorily and supplying power to an infinite bus when the prime mover failed to supply any mechanical power. The machine then runs as a

(a) Differential compound motor with speed reversed.
(b) Cumulative compound motor with speed reversed.
(e) Differential compound motor with direction of speed as before.
(d) Cumulative compound motor with direction of speed as before

105. A motor is run successively as a shunt motor, differential compound motor and cumulative compound motor with the same no-load speed. They can be arranged in the following order of decreasing full-load speeds.

(a) Cumulative compound, shunt, differential compound.
(b) Differential compound, shunt, cumulative compound.
(c) Shunt, differential compound, cumulative compound.
(d) Cumulative compound, differential compound, shunt.

106. Compensating windings in DC motors are

(a) Series excited for cancellation of armature reaction ampere turns at any load.
(b) Series excited so that armature reaction is aided at any load.
(c) Shunt excited to cancel armature reaction at any load.
(d) Shunt excited to aid the main poles at any load.

107. Which one of the following types of DC motors is the most suitable for traction purposes?

(a) Series motor
(b) Shunt motor
(c) Differential compound
(d) Cumulative compound.

108. A shunt motor may have risen mechanical characteristic due to

(a) Very high field circuit resistance.
(b) Very high armature circuit resistance.
(c) Very low demagnetizing armature reaction.
(d) Very high demagnetizing armature reaction.

109. For non-reversing DC drives it is preferable to employ
   (a) Plugging.
   (b) Dynamic braking with self-excitation.
   (c) Regenerative braking.
   (d) Dynamic braking with separate excitation.

110. In Word-Leonard method of speed control the direction of rotation of the motor is reversed usually by
   (a) reversing the connections of the generator armature terminals.
   (b) Reversing the connections of the generator field terminals.
   (c) Reversing the connections of the motor armature terminals.
   (d) Reversing the connections of the motor field terminals.

111. What counters commutation in a DC machine?
   (a) Brush resistance
   (b) Interpoles
   (c) Coil leakage inductance
   (d) Armature reaction.

112. Two DC series motors connected in series are driving the same mechanical load. If the motors are now connected in parallel the speed becomes.
   (a) Slightly more than half
   (b) Slightly less than double
   (c) Slightly less than half
   (d) Slightly more than double.

113. A DC shunt motor is running at rated speed with rated excitation, rated voltage and with on additional resistance in the armature circuit. Speeds less than the rated speed can be achieved by
   (a) Decreasing the armature circuit resistance and decreasing the field excitation.
   (b) Decreasing the field excitation and increasing the supply voltage,
   (c) Increasing the supply voltage and decreasing the armature circuit resistance.
   (d) Reducing the supply voltage and increasing the field excitation.

114. Why is it necessary to provide compensating winding in a DC motor?
   (a) To prevent a large speed drop.
   (b) To help achieve good commutation.
   (c) To reduce the main field ampere-turns.
   (d) To prevent commutator flash-over upon sudden change in load.

115. A d.c. series motor is running at rated speed with rated excitation. The motor has two resistances $R_1$, and $R_2$ connected across the armature and the field respectively. Speeds above the rated speed can be achieved by

(a) Increasing Ra only

(b) Decreasing $R_1$ only

(c) Increasing R, and decreasing Ra

(d) Decreasing $R_1$ and increasing Ra

116. D.C. motors should be stopped by opening the line switches and not by forcing the starter handle back to the off position because

(a) Heavy sparking occurs at the brushes.

(b) Heavy sparking occurs at the first stud of the starting resistance steps.

(c) Only (a).

(d) Both of the (a) and (b).

117. For DC shunt motor, speed control by armature resistance variation is best suited for

(a) Variable torque drive
(b) Constant power drive
(c) Constant torque drive
(d) Variable power drive

118. A DC series motor is driving a load with a diverter connected across If the diverter resistance is decreased, the speed of the motor

(a) Remains unchanged
(b) Increases
(c) Decreases
(d) Becomes zero

119. The most economic method of electrical braking is

(a) Plugging.

(b) Dynamic braking with self-excitation.

(c) Regenerative braking.

(d) Dynamic braking with separate excitation.

120. For DC shunt motor, speed control by armature resistance variation is best suited for

(a) Constant torque drive
(b) Variable torque drive
(c) constant power drive
(d) variable power drive.

121. A DC series motor is running with a diverter connected across its field winding. If the diverter resistance is increased then the speed of the motor

(a) Remains unchanged
(b) Becomes excessively high
(c) Decreases
(d) Increases.

122. Plugging of DC motors is normally executed by

(a) Connecting a resistance across the armature.
(b) Reversing both armature and field polarity.
(c) Reversing the armature polarity.
(d) Reversing the field polarity.

123. D.C. motor starters are used

(a) To limit the starting current.
(b) To increase the starting torque.
(c) Both (a) and (b).
(d) None of the above.

124. A DC series motor is running at rated speed without any additional resistance in series. If an additional resistance is placed in series the speed of the motor

(a) Remains unchanged
(b) Oscillates around the rated speed
(c) Decreases
(d) Increases

125. For DC shunt motors the field excitation is kept at maximum value during starting to

(a) Prevent voltage dip in the supply mains.
(b) Increase acceleration time.
(c) Decrease starting torque.
(d) Reduce armature heating.

126. A DC shunt is driving a mechanical load at rated voltage and rated excitation. If the load torque becomes double then the speed of the motor

(a) Becomes half
(b) Increases slightly
(c) Decreases slightly
(d) Becomes double.

127. A DC shunt motor is running at rated speed with rated supply voltage. If the supply voltage is halved, then the speed of the motor becomes.

(a) Half of the rated speed.
(b) Slightly less than the rated speed
(c) Slightly more than the rated speed.
(d) Double the rated speed.

128. If the field circuit of a DC shunt motor running at rated speed gets open-circuited, then immediately after this the speed of the motor would tend to

(a) Increase. (b) Decrease.

(c) Oscillate around the rated speed. (d) Remain unchanged.

129. Direct-on-line starters are not suitable for starting large DC motors because

(a) Of variable torque drive.

(b) Large voltage drop may occur in the supply mains.

(c) The motor may not start.

(d) The motor may run away.

130. A DC shunt motor is driving a constant torque load with rated excitation. If the field current is halved then the speed of the motor

(a) Becomes slightly less than double (b) Becomes double

(c) Becomes slightly more than half (d) Becomes half

131. A DC shunt motor has two additional resistances R1 and R in the field circuit and armature circuit respectively. The starting armature current can be kept to a minimum by keeping

(a) $R_1$ minimum and R2 minimum. (b) R maximum and R2 maximum.

(c) $R_1$ minimum and R2 maximum. (d) $R_1$ maximum and R2 minimum.

132. A DC shunt motor is driving a constant torque load without any additional resistance in the armature circuit. If an additional resistance is placed in the armature circuit then the speed of the motor

(a) Remains unchanged (b) Becomes zero

(c) Decreases (d) Increases.

133. Three-point starters of DC shunt motors are not used in applications where speed variation by field flux control is required because

(a) The motor may run away.

(b) The motor may stop both at very high and at very low speeds.

(c) The motor may stop at very low speeds,

(d) The motor may stop at very high speeds.

134. In Swinburne's method for the determination of efficiency of a DC machine

(a) Both the no-load losses and the copper losses are measured.

(b) The no-load losses are calculated and the copper losses are measured.

(c) The no-load losses are measured and the copper losses are calculated.

(d) Both the no-load losses and the copper losses are calculated.

135. The field's efficiency test of DC series motors overcome the difficulty of obtaining readings at relatively light loads by connecting

(a) The series field of the motor in series with the generator armature.

(b) The armature of the generator in series with the motor.

(c) The armature of the motor in series with the generator.

(d) The series field of the generator in series with the motor armature.

136. The brake test for the determination of efficiency of a DC machine is

(a) A direct method
(b) A regenerative method
(c) An indirect method
(d) None of the above.

137. The core losses in a DC machine occur due to

(a) Hysteresis only.

(b) Both hysteresis and eddy currents.

(c) Armature current.

(d) Eddy current only.

138. Commutation conditions at full-load for a large DC machine can be checked by

(a) The Swinburne's test
(b) The Hopkinson's test
(c) The brake test
(d) None of the above.

139. The core losses in a DC machine occur in

(a) Both the armature and the pole face
(b) The armature only.
(c) The Yoke only
(d) The pole faces only.

140. The efficiency of a d.c. machine is maximum when the variable losses are equal to

(a) The square of the constant losses.
(b) The constant losses.
(c) Zero.
(d) The square root of the constant losses.

141. In the Kapp's modification of Hopkinson's efficiency test

(a) The power output of the generator is dissipated in a resistor

(b) The power losses in the two machines are supplied mechanically.

(c) The power losses in the two machines are supplied electrically.

(d) The two machines are mechanically decoupled.

142. The hysteresis loss per unit weight is

(a) Proportional to the thickness of laminations.

(b) Inversely proportional to the flux density.

(c) Inversely proportional to the frequency.

(d) Proportional to the frequency.

143. Ohmic losses in a DC machine occur in

(a) The brush contacts only

(b) The armature winding only

(c) The armature winding, the field winding and also in the brush contact.

(d) The field winding only.

144. If the thickness of laminations is increased, then

(a) The hysteresis loss increases

(b) The hysteresis loss decreases

(c) The eddy current loss increases

(d) The eddy current loas decreases

145. The inductor of a three-wire generator should be iron-cored

(a) Reduce the current through it.

(b) Increase the current through it.

(c) Reduce the voltage ripple in the output circuit.

(d) Improve the voltage regulation of the generator

146. Consider the following statements on cross-field generators

(i) To increase amplification factor compensating winding should be provided

(ii) The reversal of speed does not reverse the output voltage.

(iii) The number of brush studs is double of that of poles.

Select correct answer from the list given below:

(a) Only (i) and (iii) are true

(b) Only (i) and (ii) are true

(c) All the statements are true

(d) Only (ii) and (ii) are true.

147. If the number of poles in the first stage of a Rototrol is 2, the number of poles in the second stage is

(a) 6

(b) 8

(c) 4

(d) 2.

148. An amplidyne has split poles to

(a) Increase efficiency.

(b) Damp out mechanical oscillations

(c) Provide space for interpoles.

(d) Increase amplification factor.

149. Rosenberg generator is a

(a) Constant current generator at high speeds and voltage polarity changes with the direction of rotation.

(b) Constant voltage generator at low speeds and voltage polarity is independent of the direction of rotation.)

(c) A constant voltage generator at low speeds and voltage polarity changes with the direction of rotation.

(d) A constant current generator at high speeds and voltage polarity is independent of the direction of rotation.

150. If the number of poles in a cross-field generator is 2, the number of interpoles is

(a) 6 (b) 2

(c) 8 (d) 4.

151. For a given torque, reducing the field turns of a de series motor

(a) Increases its speed but armature current remains the same

(b) Decreases its speed, demanding less armature current remains the same

(c) Decreases its speed, demanding less armature current

(d) Increases its speed, demanding more armature current.

152. Mark the correct statement below in respect of armature control of a DC shunt motor

(a) Increasing armature voltage increases motor speed and also the full-load torque capability of the motor.

(b) Increasing armature voltage decreases motor speed while the full-load torque capability is unaffected.

(c) Increasing armature voltage (with fixed current) increases motor speed while the full- load torque capability of the motor remains unaffected.

(d) Increasing armature voltage decreases motor speed and also the full-load torque capability of the motor.

153. In Hopkinson's test on two identical DC shunt motors

(a) Iron loss in the motoring machine is more than that in the generating machine

(b) Only stray-load components of iron loss in both machines are equal

(c) Iron loss in the generating machine is more than that in the motoring machine

(d) Iron losses in both machines are equal?

154. Consider the following statements in respect of compensating windings in DC motors

(i) Compensating windings are connected in series with the armature.

(ii) Compensating windings produce mmf in the same direction as armature mmf

(iii) Compensating windings aid commutation.

Mark the correct answer below:

(a) Statements (i) and (iii) are true but (ii) is false

(b) Statements (i), (ii) and (iii) are all true

(c) Statement (i) is true but (ii) and (iii) are false

(d) Statements (i) and (ii) are true but (iii) is false

155. A de series motor should not be run at light/no-load, because

(a) It will stall

(b) It will draw a dangerously large current

(c) It will draw a dangerously high current and run at a dangerously high speed

(d) It will run at a dangerously high speed.

156. In a de machine the armature reaction and the inductance of the commutating coils result in

(a) linear commutation (b) Over-commutation

(c) Under-commutation.

157. In series-parallel control of a DC series motor the total field turns are N, then

(a) $AT_{parallel} = AT_{series}$ (b) $AT_{parallel} = 0.25*(AT_{series})$

(c) AT parallel = 2AT series (d) $AT_{parallel} = 0.5*(AT_{series})$

## ANSWERS

| | | | | | | | |
|---|---|---|---|---|---|---|---|
| 1 | b | 41 | c | 81 | c | 121 | c |
| 2 | c | 42 | c | 82 | c | 122 | c |
| 3 | b | 43 | a | 83 | d | 123 | a |
| 4 | a | 44 | a | 84 | c | 124 | c |
| 5 | c | 45 | c | 85 | a | 125 | d |
| 6 | a | 46 | b | 86 | c | 126 | c |
| 7 | b | 47 | b | 87 | a | 127 | b |
| 8 | a | 48 | c | 88 | d | 128 | a |
| 9 | c | 49 | b | 89 | c | 129 | b |
| 10 | b | 50 | a | 90 | c | 130 | a |
| 11 | b | 51 | c | 91 | a | 131 | c |
| 12 | a | 52 | a | 92 | d | 132 | c |
| 13 | a | 53 | d | 93 | c | 133 | d |
| 14 | c | 54 | b | 94 | a | 134 | c |
| 15 | c | 55 | c | 95 | a | 135 | d |
| 16 | c | 56 | d | 96 | d | 136 | a |
| 17 | a | 57 | b | 97 | c | 137 | b |
| 18 | a | 58 | c | 98 | c | 138 | b |
| 19 | a | 59 | d | 99 | c | 139 | a |
| 20 | c | 60 | c | 100 | c | 140 | b |
| 21 | b | 61 | c | 101 | c | 141 | c |
| 22 | c | 62 | a | 102 | c | 142 | d |
| 23 | b | 63 | c | 103 | c | 143 | c |
| 24 | c | 64 | c | 104 | c | 144 | c |
| 25 | d | 65 | b | 105 | b | 145 | a |
| 26 | a | 66 | b | 106 | a | 146 | c |
| 27 | b | 67 | a | 107 | b | 147 | c |
| 28 | c | 68 | a | 108 | d | 148 | c |
| 29 | b | 69 | d | 109 | d | 149 | d |
| 30 | c | 70 | a | 110 | b | 150 | d |
| 31 | a | 71 | c | 111 | c | 151 | d |
| 32 | b | 72 | c | 112 | d | 152 | c |
| 33 | d | 73 | a | 113 | d | 153 | c |
| 34 | b | 74 | b | 114 | d | 154 | a |
| 35 | c | 75 | c | 115 | c | 155 | d |
| 36 | d | 76 | a | 116 | b | 156 | c |
| 37 | d | 77 | c | 117 | c | 157 | a |
| 38 | c | 78 | a | 118 | c | | |
| 39 | c | 79 | c | 119 | c | | |
| 40 | a | 80 | a | 120 | c | | |

## III. Synchronous Machines

1. The voltage regulation of a synchronous generator with two types of load Inductive load is

| **Inductive Load** | **Capacitive Load** |
|---|---|
| (a) positive, zero or negative | negative |
| (b) positive | positive, zero or negative |
| (c) negative | positive |
| (d) positive | positive or negative |

2. The rotor of a high-speed turbo-alternator is made up solid steel forging for

   (a) Reduction bearing friction
   (b) High mechanical stress
   (c) Economy
   (d) Reduction of eddy current loss

3. Synchronous motor speed is controlled by varying

   (a) Supply voltage.
   (b) Supply frequency only.
   (c) Both supply voltage and frequency.
   (d) Field excitation.

4. The fifth space harmonic in the m.m.f. developed by balanced fundamental frequency armature currents rotates at x-times the synchronous speed with respect to the poles while

   (a) x = 1
   (b) x = 6/5
   (c) x = 7/5
   (d) x = 4/5

5. Voltage regulation of a synchronous generator calculated by impedance method is synchronous

   (a) Nearly accurate because the generator is normally operated with an unsaturated magnetic circuit.
   (b) Higher than actual because of saturation of magnetic circuit.
   (c) Lower than actual because of saturation of magnetic circuit.
   (d) Nearly accurate because it takes account of magnetic saturation.

6. In a generating synchronous machine carrying load (usual symbols are used)

   (a) E lags V by angle $\delta$
   (b) E and V are in phase.
   (c) E leads V by angle $\delta$
   (d) E and V are in phase opposite

7. A coil of 150° pitch has third harmonic pitch factor as

   (a) sin 225°
   (b) cos 45
   (c) sin 45°
   (d) cos 225°

8. The rotors of a high-speed turbo-alternator and a low-speed hydel generator are
   (a) Both salient-pole
   (b) Cylindrical and salient-pole respectively.
   (c) Saline pole and cylindrical respectively
   (d) Both cylindrical
9. The following space harmonics are absent in the m.m.f. of a 3-phase synchronous machine produced by balanced sinusoidal current in the armature
   (a) 7, 13, 19, 25 etc.
   (b) 3, 5, 7, 9, 11, 13, 15, 17, 19 etc.
   (c) 3, 9, 15, 21 etc.
   (d) 5, 11, 17, 23 etc.
10. A synchronous motor is floating on infinite mains at no-load. Its excitation is now increased
    (a) It will draw zero power factor lagging current.
    (b) It will draw unity power factor current.
    (c) It will not draw any current.
    (d) It will draw zero power factor leading current
11. A synchronous generator is operating at constant load while its excitation is adjusted to give unity pf current. If the excitation is now increased, the power factor will
    (a) Lag (b) Lead
    (c) Remain at unity (d) Become zero.
12. To eliminate $r^{th}$ harmonic from the induced e.m.f. in a phase of a synchronous machine, the pitch of the coils must be
    (a) r/(r+1)th fraction of full pitch (b) (2r-1)/r fraction of full pitch
    (c) 2r/(r+1)th fraction of full pitch (d) (r-1)/rth fraction of full pitch
13. The seventh space harmonic in the m.m.f. developed by balanced fundamental frequency armature currents rotates at x-times the synchronous speed with respect to the poles, while
    (a) x=6/7 (b) x=8/7
    (c) x=5/7 (d) x=1

14. The frequency of voltage generated in an alternator is
    (a) Inversely proportional to its number of poles and speed of rotation.
    (b) Directly proportional to its number of poles and inversely proportional to its speed of rotation.
    (c) Directly proportional to its number of poles and speed of rotation.
    (d) Inversely proportional to its number of poles and directly proportional to its speed of rotation.
15. The salient-pole rotors are not used for high-speed turbo-alternators because of
    (a) Harmful mechanical oscillations.
    (b) High centrifugal force and windage loss.
    (c) Large eddy loss
    (d) Excessive bearing friction.
16. Armature reaction AT of a synchronous generator at rated voltage zero power factor lagging is
    (a) Cross magnetizing
    (b) Magnetizing
    (c) Demagnetizing
    (d) Both magnetizing and cross-magnetizing.
17. If diameter and length of the stator bore of a high-speed turbo-alternator and a low-speed hydel generator of identical rating are compared, then,
    (a) The turbo-alternator has larger diameter and smaller length.
    (b) The turbo-alternator has smaller diameter and larger length.
    (c) The turbo-alternator has larger diameter and larger length.
    (d) The turbo-alternator has smaller diameter and smaller length.
18. For a uniformly distributed winding with a phase spread of B degrees, the distribution factor at fundamental frequency is
    (a) $\dfrac{\sin\frac{\beta}{2}}{\beta}\times\dfrac{360}{\pi}$ (b) $\dfrac{2\sin\frac{\beta}{2}}{\beta}$
    (c) $\dfrac{\sin\beta}{\beta}$ (d) $\dfrac{\sin\beta}{\beta}\times\dfrac{180}{\pi}$

18. The excitation of a turbo generator feeding lagging current to infinite busbars is increased. Which of the following events will take place?
    (a) There would be no change.
    (b) The generator would feed more real power.
    (c) The generator would feed more leading kVAR.
    (d) The generator would feed more lagging kVAR.

19. For the measurement of negative sequence reactance two terminals of an unloaded excited alternator running at rated speed are shorted through an ammeter and the current coil of a wattmeter. The pressure coil of the watt- meter is connected across the voltage V which is the voltage between the free terminal and one of the shorted terminals of the alternator. If ammeter reads I and wattmeter reads W, then negative sequence impedance Z2 per phase and negative sequence reactance X2 per phase are determined as:
    (a) $Z_2 = \frac{V}{I}, X_2 = \sqrt{Z_2^2 - \left(\frac{W}{I^2}\right)^2}$
    (b) $Z_2 = \frac{V}{\sqrt{3}I}, X_2 = \sqrt{Z_2^2 - \left(\frac{W}{\sqrt{3}I^2}\right)^2}$
    (c) $Z_2 = \frac{V}{\sqrt{3}I}, X_2 = \frac{W}{\sqrt{3}I^2}$
    (d) $Z_2 = \frac{V}{I}, X_2 = \frac{W}{I^2}$

20. In a synchronous generator
    (a) The open circuit voltage leads the terminal voltage by an angle known as overlap angle.
    (b) The open circuit voltage lags the terminal voltage by an angle known as overlap angle.
    (c) The open circuit voltage lags the terminal voltage by an angle known as power angle.
    (d) The open circuit voltage leads the terminal voltage by an angle known as power angle.

21. ZPFC (Zero Power Factor Characteristic) of a synchronous generator is obtained by performing the following test:
    (a) Generator is loaded to give varying zero pf leading current while its excitation is adjusted to give the rated terminal voltage
    (b) Generator is loaded to give rated zero pf lagging current and its excitation is adjusted to vary the terminal voltage.
    (c) Generator is loaded to give varying zero pf lagging current while its excitation is adjusted to give the rated terminal voltage.
    (d) Generator is loaded to give rated zero pf leading current and its excitation is adjusted to vary the terminal voltage.

22. In calculation of regulation of a synchronous generator
    (a) While e.m.f. method gives optimistic value, m.m.f., method gives pessimistic value.
    (b) Both e.m.f. and m.m.f. methods give pessimistic values.
    (c) while e.m.f. method gives pessimistic values m.m.f. method gives optimistic value.
    (d) Both e.m.f. and m.m.f. methods give optimistic value.
23. In a synchronous generator
    (a) The field mmf leads the airgap flux and the airgap flux leads the armature m.m.f.
    (b) The field m.m.f. lags the airgap flux and the airgap flux lags the armature m.m.f.
    (c) The field mmf lags the airgap flux and the airgap flux leads the armature m.m.f.
    (d) The field leads the airgap flux and the airgap flux lags the armature m.m.f.
24. By slip test on a three-phase alternator the maximum and minimum impressed voltages per phase are found to be V, and V, respectively whereas the maximum and minimum phase currents are found to be I, and 12 respectively. The value of the direct axis synchronous reactance Xa and quadrature axis synchronous reactance X, are determined as
    (a) $X_d = \frac{V_1 + V_2}{I_1 - I_2}, X_q = \frac{V_1 - V_2}{I_1 + I_2}$
    (b) $X_d = \frac{V_1}{I_1 - I_2}, X_q = \frac{V_2}{I_1 + I_2}$
    (c) $X_d = \frac{V_1}{I_2}, X_q = \frac{V_2}{I_1}$
    (d) $X_d = \frac{V_1}{I_1}, X_q = \frac{V_2}{I_2}$
25. Consider the following statements regarding synchronous machines
    (i) Potier reactance is slightly higher than leakage reactance because of excessive saturation of the field poles during zero power factor test.
    (ii) Load characteristic of a synchronous generator represents terminal voltage as a function of field current of constant load current and power factor.
    (iii) The regulation curve of a synchronous generator shows the relationship of field current with the load current at constant power factor or with power factor at constant load current.
    (a) (ii) and (iii) are correct.

(b) (i), (ii) and (iii) are correct.

(c) (i), (ii) and (iii) are correct.

(d) all the above statements are correct.

26. A synchronous generator is feeding a certain amount of real power to infinite busbars and has normal excitation. With the real power remaining fixed and the excitation increased/decreased, the line current would

| Excitation increase | Excitation decrease |
|---|---|
| (a) decrease | increase |
| (b) decrease | decrease |
| (c) increase | increase |
| (d) increase | decrease |

27. Synchronous reactance is

(a) The same as armature leakage reactance

(b) The difference of armature leakage reactance and reactance equivalent of armature reaction.

(c) The sum of armature leakage reactance and reactance equivalent of armature reaction.

(d) the reactance equivalent of armature reaction.

28. Short-circuit ratio of a synchronous machine is defined as the ratio of

(a) Field current required to produce rated voltage on OC and field current required to produce rated armature current on SC.

(b) Field current required to produce rated voltage on full-load and field current required to produce rated current on SC.

(c) Field current required to produce rated voltage on full-load and field current required to produce rated voltage on SC.

(d) Field current required to produce rated voltage on SC and field current required to produce rated current on SC.

29. The phasor diagram by ASA method gives

(a) Considerably erroneous result for regulation but reliable result for power angle.

(b) Reliable result for regulation but considerably erroneous result for power angle.

(c) Considerably erroneous results for both regulation and power angle of a salient-pole synchronous generator.

(d) Reliable results for both regulation and power angle of a salient-pole synchronous generator.

30. A synchronous generator is rated 10 MVA, 11 kV, 85 pf, 50 Hz. What is the significance of. 85 pf?

(a) Load of pf higher than 85 cannot be supplied.

(b) It is not any significance.

(c) The prime mover power rating is 8.5 MW.

(d) Load of pf lower than 85 cannot be supplied.

31. A synchronous motor is operating at constant load while its excitation is adjusted to give unity pf current. If the excitation is now increased, the power factor will

(a) remain at unity (b) Become zero

(c) Lead (d) Lag.

32. Armature reaction m.m.f. and leakage reactance of a synchronous machine are determined by

(a) Open circuit and short circuit tests.

(b) Zero power factor test only.

(c) Open circuit test only.

(d) Open circuit and zero power

33. Synchronous reactance, Xs (adjusted) is obtained as (standard symbols are used)

(a) $\frac{V_t(rated)}{I_{SC}} / I_f$ ; where $I_f$ is corresponding to $V_t$ (rated) on OCC

(b) $\frac{\frac{V_t(rated)}{\sqrt{3}}}{I_{SC}} / I_f$ ; where $I_f$ is corresponding to $V_t$ (rated) on OCC

(c) $\frac{\frac{V_t(rated)}{\sqrt{3}}}{I_{SC}} / I_f$ ; where $I_f$ is corresponding to $V_t$ (rated) on full load

(d) $\frac{\frac{V_{OC}}{\sqrt{3}}}{I_{SC}} / I_f$ ; where $I_f$ is corresponding to $V_t$ (rated) on full load

34. In a synchronous machine, the voltage induced by armature reaction flux acts like
    (a) The voltage drops in resistance.
    (b) The voltage drops in inductive impedance, reactance
    (c) The voltage drops in inductive
    (d) The voltage drops in capacitive reactance
35. In a motoring synchronous machine (usual symbols are used)
    (a) E lags V by angle δ.
    (b) E and V are in phase opposition.
    (c) E leads V by angle δ.
    (d) E and V are in phase.
36. If armature current in a generating synchronous machine is in phase with the excitation e.m.f., the armature m.m.f. vector
    (a) Is in a direction opposite to the field m.m.f. vector.
    (b) Leads the field m.m.f. vector by $90^0$
    (c) Lags behind the field m.m.f. vector by $90^0$
    (d) Is in the same direction as the field m.m.f. vector
37. From the current oscillograms of sudden symmetrical short circuit of an unloaded three-phase alternator, the peak value of the alternating component of current at the instant of short circuit is $I_1$, and the peak value of the steady- state current is $I_2$, while E is the r.m.s. voltage per phase before short circuit. Then the direct-axis synchronous reactance $X_d$ and direct-axis sub transient reactance $X_d''$ are determined by the following expressions.

    (a) $X_d = \dfrac{E}{I_1 + I_2}$, $X_d'' = \dfrac{E}{I_1 - I_2}$  (b) $X_d = \dfrac{\sqrt{2}E}{I_1 - I_2}$, $X_d'' = \dfrac{\sqrt{2}E}{I_1 + I_2}$

    (c) $X_d = \dfrac{\sqrt{2}E}{I_2}$, $X_d'' = \dfrac{\sqrt{2}E}{I_1}$  (d) $X_d = \dfrac{\sqrt{2}E}{I_1}$, $X_d'' = \dfrac{\sqrt{2}E}{I_2}$
38. In a three-phase synchronous motor
    (a) The field m.m.f. leads the airgap flux and the airgap flux lags the armature m.m.f.
    (b) The field m.m.f. lags the airgap flux and the airgap flux lags the armature m.m.f.
    (c) The field m.m.f. leads the airgap flux and the airgap flux leads the armature m.m.f.

(d) The field m.m.f. lags the airgap flux and the airgap flux leads the armature m.m.f.

39. In a synchronous machine, the induced e.m.f. phasor

(a) Is in phase with the flux phasor.

(b) Leads the flux phasor by 90°.

(c) Is in phase opposition to the flux phasor.

(d) Lags behind the flux phasor by 90°.

40. Synchronous motor speed is controlled by varying

(a) Supply voltage

(b) Both supply voltage and frequency.

(c) Supply frequency only.

(d) Field excitation.

41. During constant load, variable excitation operation of a synchronous machine (usual symbols are used)

(a) $E_f \sin\delta = \dfrac{P_e}{V_t x_S}$ (b) $E_f \cos\delta = \dfrac{P_e}{V_t x_S}$

(c) $E_f \sin\delta = \dfrac{P_e x_S}{V_t}$ (d) $E_f \cos\delta = \dfrac{P_e x_S}{V_t}$

42. A synchronous generator is feeding power to infinite busbars at unity power factor. Its excitation is now increased. It will feed:

(a) Less power at unity power factor.

(b) More power at unity power factor.

(c) The same power but at a lagging power factor.

(d) The same power but at a leading power factor.

43. Which one of the following statements regarding a three-phase hysteresis motor is correct?

(a) Both eddy-current and hysteresis torque are effective during starting while only hysteresis torque is effective during running.

(b) Both eddy current and hysteresis torques are effective during starting as well as running.

(c) Only hysteresis torque is effective during starting as well as running.

(d) Only eddy current torque is effective during starting while only hysteresis torque is effective during running

45. For maximum power input and maximum power output of a 3-phase synchronous motor working on constant excitation across infinite bus, its power angle $\delta$ is related to its internal angle $\gamma$ as

(a) $\delta=\gamma$ and $\delta= 180^0-\gamma$ respectively.

(b) $\delta= 180^0-\gamma$ and $\delta=\gamma$ respectively.

(c) $\delta= 180-2\gamma$ and $\delta= \gamma/2$ respectively.

(d) $\delta= \gamma/2$ and $\delta= 180-2\gamma$ respectively.

46. During constant load, variable excitation operation of a synchronous machine ($\delta$= power angle, $\varphi$= pf angle)

(a) $E_f \cos\delta$ = constant, $I_a \cos\phi$ = constant

(b) $E_f \cos\delta$ = constant, $I_a \sin\phi$ = constant

(c) $E_f \sin\delta$ = constant, $I_a \sin\phi$ = constant

(d) $E_f \sin\delta$ = constant, $I_a \cos\phi$ = constant

47. Armature reaction AT of a synchronous generator under short circuit-condition is

(a) Demagnetizing.

(b) Magnetizing.

(c) Cross-magnetizing

(d) Both magnetizing and cross-magnetizing

48. A synchronous motor is operated from a bus voltage of 1.0 PU and is drawing 1.0 PU zero power factor leading current. Its synchronous reactance is 0.5 PU. The excitation e.m.f. of the motor will be

(a) 20 (b) 0.5

(c) 1.5 (d) 1.0

49. The sub transient reactance of a synchronous generator is much less than its synchronous reactance in the initial phase of short-circuit from no-load as

(a) Demagnetizing ampere-turns appear because of induced field and damper winding currents as the airgap flux cannot change instantaneously from its no-load value.

(b) Magnetizing ampere-turns appear because of induced field and damper winding currents as the air-gap flux cannot change instantaneously from its no-load value.

(c) Demagnetizing ampere-turns appear because of induced field and damper winding currents as the air-gap flux must reduce instantaneously from its no-load value.

50. A synchronous motor is running from busbars, and has a torque angle of $\delta = 15$. The bus voltage and frequency are reduced by 10% while field current and load torque are held constant (all losses are negligible). The new torque angle will

(a) Remain the same. (b) The motor will lose synchronism

(c) Increase slightly. (d) Decrease slightly.

51. Upon the occurrence of a 3-phase short-circuit, DC off-set currents appear in the stator phases

(a) Which cause fundamental frequency currents in the field winding, which in turn induce second harmonic currents in the stator winding

(b) Which cause second harmonic frequency currents in the field winding, which in turn induce fundamental frequency currents in the stator winding.

(c) Which cause fundamental frequency currents in the field winding, which in turn induce harmonic current in the stator winding.

(d) Which cause second harmonic frequency currents in the field winding, which in turn induce fundamental frequency currents in the stator winding.

52. $X_d$=d-axis synchronous reactance; $X_q$=q-axis synchronous reactance In a salient pole synchronous machine

(a) $X_q < X_d$ (b) $X_q > X_d$

(c) $X_q = X_d$ (d) $X_q = 0$

53. Damper windings of a synchronous machine are

(a) Short-circuited copper bars placed in pole shoes, which prevent the rotor from running at super synchronous speed.

(b) Short-circuited copper bars placed in pole shoes, which help to reduce rotor oscillation about the operating point.

(c) Short-circuited copper bars placed on the rotor in the interpolar regions, which help reduce rotor oscillation about the operating point.

54. Potier's method uses OCC and ZPFC to yield information about

(a) Field current equivalent of armature reaction only.

(b) Leakage reactance and field current equivalent of armature reaction.

(c) Synchronous reactance.

(d) Leakage reactance only.

55. The pole shoes of a salient pole synchronous machine are so shaped that air-gap is least in the middle and increases somewhat towards the edges. This type of construction

(a) Facilitates machining of pole shoes.

(b) Eliminates a particular harmonic.

(c) Reduces the harmonic content of the induced voltage wave.

(d) Increases flux/pole for a given excitation.

56. If V be the terminal voltage per phase, Eo be the open-circuit voltage per phase and $\gamma$ be the internal angle of a 3-phase synchronous motor, its power factor will always be lagging if

(a) $E_o < V$

(b) $Eo < V \sin \gamma$

(c) $Eo \sin \gamma < V$

(d) $Eo \sin y < V$.

57. Consider the following statements

(i) A synchronous motor has no starting torque but when started it always runs at a fixed speed.

(ii) Neglecting residual magnetism, both cylindrical-rotor and salient pole synchronous motors produce some mechanical power even though the field is unexcited.

(iii) A single-phase reluctance motor is not self-starting even if paths for eddy currents are provided in the rotor.

(iv) A single-phase hysteresis motor is self-starting.

(a) Only (i) is correct.

(b) Only (i) and (iii) are correct

(c) Only (i), (iii) and (iv) are correct.

(d) All the above statements are correct.

58. The operation of a 3-phase synchronous motor working on constant excitation across infinite bus cannot be stable if the following relationship between power angle $\delta$ and internal angle $\gamma$ exists

(a) $\delta > \gamma$

(b) $2\delta > \gamma$

(c) $2\delta < \gamma$

(d) $\delta < \gamma$

59. A synchronous generator is loaded and the armature current lags behind the excitation e.m.f. by angle v. The electrical angle between the field axis and the axis of the armature reaction field will be

(a) $\psi$ (b) 90

(c) 90° - $\psi$ (d) 90+ $\psi$

60. As observed from the stator of a synchronous generator, the armature reaction m.m.f. phasor

(a) Is in phase opposition to the armature current phasor

(b) Leads the armature current phasor by 90°.

(c) Is in phase with the armature current phasor.

(d) Lags behind the armature current phasor by 90".

61. Synchronous generator voltage obtained by the synchronous impedance method as

(a) Nearly accurate as the generator is normally operated in the unsaturated region of magnetization.

(b) Lower than actual as it does not account for magnetic saturation.

(c) Higher than actual as it does not account for magnetic saturation.

(d) Nearly accurate as it accounts for magnetic saturation.

62. A 3-phase synchronous motor driving a constant load torque draws power from the infinite bus at a leading power factor. If the excitation is increased

(a) Both power angle and power factor decrease.

(b) The power angle decreases while power factor increases.

(c) The power angle increases while power factor decreases.

(d) Both power angle and power factor increase.

63. In a salient pole synchronous machine (usual symbols are used)

(a) $E_f = V_t \sin \delta \pm I_d X_d$ (+ for generator, for motor)

(b) $E_f = V_t \sin \delta \pm I_d X_d$ (+ for motor, - for generator)

(c) $E_f = V_t \sin \delta \pm I_d X_d$ (+ for gen, - for motor)

(d) $E_f = V_t \sin \delta \pm I_d X_d$ (- for generator, - for motor)

64. A 3-phase synchronous motor with constant excitation is driving a certain load drawing electrical power from infinite bus at leading power factor. If the shaft load decreases

(a) The power angle increases while power factor decreases.

(b) Both power angle and power factor decrease.

(c) The power angle decreases while power factor increases.

(d) Both power angle and power factor increase.

65. For constant load operation of a synchronous machine, the minimum excitation e.m.f. is

(a) $E_{f(\min)} = \frac{P_e x_S}{V_t}$ at δ=90⁰ (b) $E_{f(\min)} = \frac{P_e V_t}{x_S}$ at δ=90⁰

(c) $E_{f(\min)} = \frac{P_e V_t}{x_S}$ at δ=0⁰ (d) $E_{f(\min)} = \frac{P_e x_S}{V_t}$ at δ=0⁰

66. If an oscillating synchronous machine is stable, the maximum phase-plane trajectory and the point at which it touches the abscissa are respectively known as

(a) Separatrix and critical point.

(b) Separatrix and saddle point.

(c) Critical curve and critical point.

(d) Critical characteristic and saddle point.

67. Let $f_s$ and $f_r$ be the frequencies of the bus and the incoming machine to be connected to the bus. Then state which one of the following is true concerning three-dark method and two-bright one-dark method of synchronization by lamps.

(a) Both give fs-fr

(b) Both give Mod(fs-fr) only

(c) The former gives fs-fr while the latter gives Mod(fs-fr)only

(d) The former gives Mod(fs-fr) only while the latter gives fs-fr

67. (a) Upon the occurrence of 3-phase short-circuit of a synchronous machine, d.c. off-set currents are setup in the three phases

(a) Which are proportional to the sine of the angle between the phase axis and d-axis. These produce a fixed axis field in the air-gap.

(b) Which are proportional to the cosine of the angle between the phase axis and d-axis. These produce a rotating field in the air-gap.

(c) Which are proportional to the sine of the angle between the phase axis and d-axis. These produce a rotating field in the air-gap.

(d) Which are proportional to the cosine of the angle between the phase axis and d-axis These produce a fixed axis field in the air-gap.

68. Compare the per pole permeance of direct and quadrature axis of a round rotor machine

(a) Quadrature axis permeance is more than direct axis permeance.

(b) Direct axis permeance is more than quadrature axis permeance.

(c) Quadrature axis permeance is zero.

(d) Direct and quadrature axes permeances are equal.

69. The steam input to a turbo generator connected to infinite busbars is increased. Which of the following events will take place?

(a) The generator will feed more real power to busbars and its power angle will decrease.

(b) The generator will feed more real power to busbars and its power angle will increase.

(c) The generator will feed more lagging kVAR to busbars but its power angle will not change.

(d) The generator will feed more leading kVAR to busbars but its power angle will not change.

70. The swing equation of a synchronous machine is given as

$$J1\frac{d^2\delta}{dt^2}+D\frac{d\delta}{dt}=P_a(\delta)$$

where $P_a(\delta)$ is the accurating power as a function of power angle δ. If H is the kinetic energy of the rotor at synchronous speed and the frequency of supply then,

(a) $J1=\dfrac{H}{4\pi f}$ (b) $J1=\dfrac{2H}{\pi f}$

(c) $J1=\dfrac{H}{\pi f}$ (d) $J1=\dfrac{H}{2\pi f}$

71. In a motoring synchronous machine as observed from the stator, the armature m.m.f. phasor

(a) Leads the armature current phasor by 90°.

(b) Lags behind the armature current phasor by 90°.

(c) Is in phase opposition to the armature current phasor.

(d) Is the phase with the armature current phasor.

72. Armature reaction AT of a synchronous motor at rated voltage zero power Factor leading is
    (a) Magnetizing
    (b) Cross-magnetizing
    (c) Demagnetizing.
    (d) Both cross-magnetizing and demagnetizing.
73. The voltage regulation of a synchronous generator with two types of loads is

| | Inductive load | Capacitive load |
|---|---|---|
| (a) | Negative | Positive. |
| (b) | Positive | Positive or negative |
| (c) | Positive | Positive, zero or negative |
| (d) | Positive, zero or negative | Negative |

74. A 3-phase alternator with constant excitation is supplying electrical power to negative. an infinite bus at lagging power factor. If steam input decreases
    (a) The power angle increases while power factor decreases.
    (b) The power angle decreases while power factor increases.
    (c) Both power angle and power factor increase.
    (d) Both power angle and power factor decrease.
75. SCR (Short Circuit Ratio) of a synchronous machine yield
    (a) $\frac{1}{X_S(adjusted)}$ (b) $\frac{1}{X_S(unsaturated)(pu)}$
    (c) $\frac{1}{X_S(unsaturated)}$ (d) $\frac{1}{X_S(adjusted)(pu)}$
76. Two alternators of capacities 1000 kVA and 1500 kVA operate in parallel. The governor mechanisms are such that they have a common no-load speed and have speed drops of 3% and 4.5% respectively on full-loads. A total load of 1000 kW will be distributed between them as
    (a) 500 kW and 500 kW respectively.
    (b) 450 kW and 550 kW respectively
    (c) 600 kW and 400 kW respectively.
    (d) 400 kW and 600 kW respectively.

77. Synchronous reactance, $X_s$ (unsaturated) is obtained as:

$V_{oc}$ = open-circuit voltage (line)

$I_{sc}$ = short-circuit current (line)

$I_f$ = field current.

(a) $\left(\dfrac{V_{OC}/\sqrt{3}}{I_{SC}} / I_f\right)$ = constant; in non-linear region of OCC

(b) $\left(\dfrac{V_{OC}/\sqrt{3}}{I_{SC}} / I_{f(\text{constant})}\right)$ ; in non-linear region of OCC

(c) $\left(\dfrac{V_{OC}}{I_{SC}} / \text{any } I_f\right)$

(d) $\left(\dfrac{V_{OC}}{I_{SC}} / I_{f(rated)}\right)$

78. For a given excitation voltage a synchronous generator connected to infinite bus will develop maximum output power and maximum electromagnetic power when the power angle δ is related to the internal angle γ as

(a) $\delta = \gamma$ and $\delta = 180 - 2\gamma$
(b) $\delta = \gamma$ and $\delta = 180 - \frac{\gamma}{2}$
(c) $\delta = \gamma$ and $\delta = 180 - \gamma$
(d) $\delta = 180 - \gamma$ and $\delta = \gamma$

79. The stator of a salient pole synchronous generator is excited from a low voltage 50 Hz supply while the rotor is made to run at a slip of 0.05. The reluctance presented to the rotating stator m.m.f.

(a) Varies sinusoidally at 50 Hz
(b) Varies sinusoidally at 52.5 Hz
(c) Varies sinusoidally at 2.5 Hz
(d) Constant

80. A synchronous motor with 5-ohm synchronous reactance (negligible resistance) draws a leading current of 10 A at 400 V. The induced e.m.f. is

(a) $400 - j\sqrt{3} \times 50$
(b) $400 - \sqrt{3} \times 50$
(c) $400 + \sqrt{3} \times 50$
(d) $400 + j\sqrt{3} \times 50$

81. A 3-phase synchronous generator with constant steam input supplies power to an infinite bus at a lagging power factor. If the excitation is increased

(a) The power angle increases while power factor decreases.

(b) The power angle decreases while power factor increases.

(c) Both power angle and power factor increase.

(d) Both power angle and power factor decrease.

82. A three-ring synchronous converter supplying a DC 3-wire system must have the secondary of its transformer connected as

(a) Zigzag only
(b) Delta only
(c) Star or zigzag
(d) Delta or star.

83. Compare the per pole permeance of direct and quadrature axes of a salient pole synchronous machine

(a) Quadrature axis permeance is zero.

(b) Direct and quadrature axis permeances are equal.

(c) Quadrature axis permeance is more than direct axis permeance.

(d) Direct axis permeance is more than quadrature axis permeance.

84. The number of slip rings of a single-phase and three-phase converters are respectively

(a) 2 and 3
(b) 1 and 3
(c) 2 and 6
(d) 1 and 6

85. In a salient pole synchronous machine (usual symbols are used) reluctance power is given by

(a) $V_{bus}^2\left(\frac{X_d + X_q}{2X_d X_q}\right)\sin 2\delta$
(b) $V_{bus}^2\left(\frac{X_d - X_q}{2X_d X_q}\right)\sin \delta$
(c) $V_{bus}^2\left(\frac{X_d - X_q}{2X_d X_q}\right)\sin 2\delta$
(d) $V_{bus}^2\left(\frac{X_d + X_q}{2X_d X_q}\right)\sin \delta$

86. If A be the armature m.m.f. in the interpolar axis of a rotary converter running as a DC generator, then the m.m.f. at the same position while running as a converter supplying the same DC load at efficiency η, is

(a) $\left(1+\frac{\pi^2}{8\eta}\right)A$
(b) $\left(1-\frac{\pi^2}{8\eta}\right)A$
(c) $\left(1+\frac{8}{\pi^2\eta}\right)A$
(d) $\left(1-\frac{8}{\pi^2\eta}\right)A$

87. The maximum electrical power input of a synchronous motor is (usual symbols are used)

(a) $\frac{V_t E_f}{X_S}$ (b) $\frac{X_S}{V_t E_f}$

(c) $\frac{{E_f}^2}{X_S}$ (d) $\frac{V_t^2}{X_S}$

88. By connecting reactance at the slip-ring leads of a rotary converter running at a leading power factor, the output DC voltage can be raised if

(a) The power factor is made less leading or lagging by decreasing the excitation

(b) The power factor is made more leading by decreasing the excitation.

(c) The power factor is made more leading by increasing the excitation.

(d) The power factor is made less leading or lagging by increasing the excitation,

89. d=direct axis, q=quadrature axis, $E_f$= excitation e.m.f. In a salient pole synchronous machine

(a) $I_d$ and $I_q$ are at $90^0$ to $E_f$

(b) $I_d$ and $I_q$ are both in phase with $E_f$

(c) $I_q$ is in phase with $E_f$ and $I_d$ is at $90^0$ to $E_f$

(d) $I_q$ is at $90^0$ to $E_f$ and $I_d$ is in phase with $E_f$

90. If H be the ratio of heating in the armature of a synchronous converter running as a converter to that while running as a DC generator with the same DC output, then the ratio of outputs as a converter to that/as a DC generator for the same armature heating is:

(a) 1/H (b) H

(c) $\sqrt{H}$ (d) $\frac{1}{\sqrt{H}}$

91. The maximum electrical power output of a synchronous generator is (usual symbols are used)

(a) $\frac{V_t^2}{X_S}$ (b) $\frac{{E_f}^2}{X_S}$

(c) $\frac{V_t E_f}{X_S}$ (d) $\frac{X_S}{V_t E_f}$

92. Which one of the following statements is true about the position of minimum copper loss in a phase of a synchronous converter?

(a) The position of minimum copper loss varies with the power factor and this position is displaced from the tapping point by the power factor angle.

(b) The position of minimum copper loss is the tapping point whatever may be the power factor

(c) The position of minimum copper loss varies with the factor and this position is displaced from the middle of the phase by the power factor angle.

(d) The position of minimum copper loss is the middle of the phase whatever may be the power factor.

93. In a synchronous machine, let

$I_{f1}$ field current for rated open-circuit voltage.

$I_{f2}$ field current for rated current on short-circuit.

The short-circuit ratio of the machine is defined as

(a) $I_{f1}/I_{f2}$ and equals the PU synchronous reactance.

(b) $I_{f1}/I_{f2}$ and equals the reciprocal of PU synchronous reactance.

(e) $I_{f2}/I_{f1}$ and equals the PU synchronous reactance.

(d) $I_{f2}/I_{f1}$ and equals the reciprocal of the PU synchronous reactance,

94. A salient pole synchronous machine can stay synchronized to the main when excitation is reduced to zero provided the load is

(a) more than $V_t^2\left(\frac{X_d+X_q}{2X_dX_q}\right)$

(b) less than $V_t^2\left(\frac{X_d-X_q}{2X_dX_q}\right)$

(c) less than $V_t^2\left(\frac{X_d+X_q}{2X_dX_q}\right)$

(d) more than $V_t^2\left(\frac{X_d-X_q}{2X_dX_q}\right)$

95. Which one of the following statements regarding the ratios of line voltages and line currents on the a.c. side to that on the d.c. side of a polyphase converter is true?

(a) The former is greater than the unity while the latter is less than unity.

(b) Both are less than unity.

(c) Both are greater than unity.

(d) The former is less than unity while the latter is greater than unity.

96. If the excitation of a salient pole motor is reduced to zero

(a) it will remain synchronized provided the load is less than a certain value.

(b) it will lose synchronism.

(c) it will remain synchronized.

(d) it will remain synchronized provided it is operating at no-load.

97. Which one of the following statements regarding the ratios of line voltages and line currents on the AC side to that on the DC side of a rotary converter is true?

(a) The former decreases while the latter increases with increase in the number of phases.

(b) Both decrease with increase in number of phases.

(c) Both increase with increase in number of phases.

(d) The former increases while the latter decreases with increase in the number of phases.

98. Two synchronous generators $G_1$ and G2 are operating in parallel and shorting the load equally. To shift part of the load from $G_2$ to G1 while keeping the frequency constant, the following actions must be taken.

(a) Raise the frequency power characteristic of G1 or lower that of $G_2$.

(b) Lower the frequency-power characteristic of G1 or raise that of G2.

(c) Lower the frequency-power characteristic of $G_1$ and raise that of G2.

(d) Raise the frequency-power characteristic of G1 and lower that of G2.

99. If the frequency of supply to a motor converter be f Hz, the number of poles of the induction motor be Pm, the number of poles of the converter be Pg, and the DC power generated be P, then the speed of the set and power transferred mechanically to the converter from the induction motor are

(a) $\frac{120f}{P_g}$ *rpm and* $\frac{P_g}{P_m + P_g} \times P$ *respectively*

(b) $\frac{120f}{P_m}$ *rpm and* $\frac{P_m}{P_m + P_g} \times P$ *respectively*

(c) $\frac{120f}{P_m + P_g}$ *rpm and* $\frac{P_m}{P_m + P_g} \times P$ *respectively*

(d) $\frac{120f}{P_m + P_g}$ *rpm and* $\frac{P_g}{P_m + P_g} \times P$ *respectively*

100. Two synchronous generators G1 and G2 are operating in parallel and are equally sharing KVAR (lagging) component of load. To shift part of the KVAR load from G2 to G1 while keeping the terminal voltage fixed, the following action must be taken

(a) Raise the field current of G1 and lower that of G2

(b) Lower the field current of G1 and raise that of G2

(c) Lower the field current of G1 or raise that of G2

(d) Raise the field current of G1 or lower that of G2

101. The synchronizing coefficient of a synchronous machine is

(a) $\left(\frac{d\delta}{dP_e}\right)$ at operating point, it increases as δ increases.

(b) $\left(\frac{d\delta}{dP_e}\right)$ at operating point, it decreases as δ increases.

(c) $\left(\frac{dP_e}{d\delta}\right)$ at operating point, it decreases as δ increases.

(d) $\left(\frac{dP_e}{d\delta}\right)$ at operating point, it increases as δ decreases.

102. Upon the occurrence of 3-phase short-circuit of a synchronous machine the symmetrical component of current gradually reduces to a steady-state value. It is because

(a) damper winding current somewhat increases while field current decays. As the latter predominates it results in a gradual reduction of the air-gap flux to a steady value.

(b) both field and damper winding currents decay resulting in gradual reduction of the air-gap flux to a steady value.

(c) damper winding current decays while field current remains constant resulting in gradual reduction of the air-gap flux to a steady value.

(d) field current decays while damper winding current remains constant resulting in gradual reduction of the air-gap flux to a steady value.

103. In "slip test" on a synchronous machine, the stator m.m.f. alignment, for maximum/minimum current drawn from mains, is

| | Minimum current | Maximum current |
|---|---|---|
| (a) | along $45^0$ to d-axis | along $45^0$ to q-axis |
| (b) | along q-axis | along d-axis |

(c) along d-axis along q-axis

(d) along 45 to q-axis along $45^0$ to d-axis.

104. The armature current upon symmetrical 3-phase short-circuit of a synchronous machine (armature resistance is negligible)

(a) has both d-axis and q-axis components.

(b) constitutes q-axis current only.

(c) constitutes d-axis current only.

(d) short-circuit current cannot be divided into d- and q-axis components.

105. The distribution factor for a 36 slot, 4 pole, 3 phase winding is:

(a) 0.960 (b) 0.892

(c) 0.403 (d) 0.709

106. In a 3-phase, star connected alternator a field current of 40 Amps, produces full load current of 200 Amps on short circuit and 1160 V on open circuit. If the resistance of the alternator is 0.5 ohm, then the value of synchronous reactance is

(a) 16.20 (b) 3.98

(c) 3.31 (d) 20 ohms

107. A 10 pole, 25 c/s alternator is directly coupled to, and is driven by a 60 c/s synchronous motor. Then the number of poles in a synchronous motor are

(a) 24 (b) 12

(c) 48 (d) 36

108. Reactive power output (lagging) of a synchronous generator is limited by

(a) armature current (b) load angle

(c) field current (d) prime mover input.

109. In a salient pole synchronous generator connected to an infinite bus-bar maximum power is delivered at a power angle where

(a) $\delta=0°$ (b) $\delta= 45°$

(c) $45 < \delta < 90°$ (d) $5 = 90$

110. At constant load, the magnitude of armature current of a synchronous motor have large values for

(a) low values of field excitation

(b) high values of field excitation

(c) both low values and high values of excitation

(d) none of the above.

## ANSWERS

| 1 | B | 26 | C | 51 | A | 76 | A | 101 | C |
|---|---|---|---|---|---|---|---|---|---|
| 2 | B | 27 | C | 52 | A | 77 | B | 102 | B |
| 3 | C | 28 | A | 53 | B | 78 | C | 103 | B |
| 4 | B | 29 | A | 54 | B | 79 | C | 104 | B |
| 5 | B | 30 | B | 55 | C | 80 | C | 105 | A |
| 6 | C | 31 | C | 56 | B | 81 | D | 106 | C |
| 7 | B | 32 | C | 57 | B | 82 | A | 107 | A |
| 8 | B | 33 | D | 58 | A | 83 | D | 108 | C |
| 9 | C | 34 | B | 59 | D | 84 | A | 109 | D |
| 10 | D | 35 | C | 60 | C | 85 | B | 110 | C |
| 11 | A | 36 | A | 61 | C | 86 | D | | |
| 12 | D | 37 | C | 62 | A | 87 | A | | |
| 13 | A | 38 | C | 63 | B | 88 | C | | |
| 14 | C | 39 | B | 64 | B | 89 | C | | |
| 15 | B | 40 | D | 65 | A | 90 | D | | |
| 16 | D | 41 | B | 66 | B | 91 | C | | |
| 17 | B | 42 | C | 67 | D | 92 | C | | |
| 18 | B | 43 | C | 68 | A | 93 | D | | |
| 19 | C | 44 | A | 69 | B | 94 | B | | |
| 20 | D | 45 | B | 70 | C | 95 | B | | |
| 21 | B | 46 | D | 71 | C | 96 | A | | |
| 22 | C | 47 | A | 72 | C | 97 | B | | |
| 23 | A | 48 | C | 73 | C | 98 | D | | |
| 24 | C | 49 | B | 74 | D | 99 | C | | |
| 25 | B | 50 | C | 75 | D | 100 | A | | |

## IV. Polyphase Induction Motor

## AND

## Other AC Motors

1. Skewing of rotor slots

(a) Decreases both stator resistance and noise.

(b) Increases both stator resistance and noise.

(c) Increases rotor resistance but decreases noise.

(d) Decreases rotor resistance but increases noise.

2. The resistance Ro of the exciting branch of the equivalent circuit of a 3-phase induction motor represents

(a) rotor copper loss. (b) stator core loss only..

(c) friction and windage losses only. (d) stator copper loss

3. If $P_G$ represents the power across the air-gap in an induction motor, then

| | Mechanical output | Rotor copper loss |
|---|---|---|
| (a) | $SP_G$ | $(1+S)P_G$ |
| (b) | $(1+S)P_G$ | $SP_G$ |
| (c) | $(1-S)P_G$ | $SP_G$ |
| (d) | $SP_G$ | $(1-S)P_G$ |

4. Mark the correct answer below as the load on an induction motor is increased up to full-load.

| PF | Slip | Efficiency |
|---|---|---|
| (a) Decreases | Increases | Increases |
| (b) Decreases | Decreases | Decreases |
| (c) Increases | Decreases | Decreases |
| (d) Increases | Increases | Increases |

5. For controlling the speed of an induction motor the frequency of supply is increased by 10%. For magnetizing current to remain the same, the supply voltagc must

(a) Remain constant (b) Be increased by 10%.

(c) Be reduced or increased by 10% (d) Be reduced by 10%.

6. A 3-phase, 50 Hz, 8-pole squirrel cage induction motor will run at a speed

(a) $>$ 750 r.p.m., but $<$ 1000 r.p.m. (b) $<$ 750 r.p.m.

(c) $>$ 1500 r.p.m. (d) $>$ 1000 r.p.m. but $<$ 1500 r.p.m.

7. Induction generators deliver power

(a) At zero power factor (b) t unity power factor

(c) At leading power factor (d) At lagging power factor

8. Complete circle diagram of a 3-phase induction motor can be constructed from

(a) Ratio of stator and equivalent rotor resistance only.

(b) No-load and blocked rotor test data and ratio of stator and equivalent rotor resistance.

(c) Blocked rotor test data only

(d) No-load test data only

9. Mark the correct statement below in respect of an induction motor (standard symbols are used)

| Power across air-gap | Rotor Copper Loss | Mechanical Power Output |
|---|---|---|
| (a) $\frac{3I_2'^2 r_2'}{S}$ | SPG | $\left(\frac{1}{S}-1\right)P_G$ |
| (b) $\frac{3I_2'^2 r_2'}{S}$ | (1-S) PG | SPG |
| (c) $\frac{3I_2'^2 r_2'}{S}$ | (1-S) PG | SPG |
| (d) $\frac{3I_2'^2 r_2'}{1-S}$ | $\left(\frac{1}{S}-1\right)P_G$ | SPG |

10. If the rotor winding of a wound rotor 3-phase induction motor is connected to balanced 3-phase supply and the stator winding is short-circuited, then the rotor will

   (a) Run at sub-synchronous speed in the direction of rotating flux.

   (b) Run at sub-synchronous speed against the direction of rotating flux.

   (c) Run synchronously with rotating flux.

   (d) Not run.

11. For maximum starting torque in an induction motor

   (a) $r_2 = 2x_2$ (b) $r_2 = 0$

   (c) $r_2 = x_2$ (d) $r_2 = 0.5x_2$

12. Induction generators run at

   (a) Very low sub-synchronous speed.

   (b) Sub-synchronous speed very near to synchronous speed.

   (c) Synchronous speed

   (d) Super synchronous speed.

13. The resistance representing mechanical output in the equivalent circuit of an induction motor as seen from the stator is

   (a) $\frac{r_2'^2}{S}$ (b) $\frac{r_2}{S}$

   (c) $r_2'\left(\frac{1}{S}-1\right)$ (d) $r_2^2\left(\frac{1}{S}-1\right)$

14. In the case of double-cage rotors

   (a) The outer cage has low resistance and the inner cage has high resistance

   (b) The outer cage has high resistance and the inner cage has low resistance.

(c) Both the cages have low resistance.

(d) Both the cages have high resistance.

15. In a 3-phase induction motor, the variable mechanical load is electrically represented by

(a) A combination of variable resistance inductance.

(b) A variable capacitance only.

(c) A variable inductance only.

(d) A variable resistance only.

16. The deep-bar rotors or double-cage rotors are used

(a) To increase pull out torque

(b) To improve efficiency

(c) To reduce rotor core-losses

(d) To increase starting torque.

17. In terms of air-gap power $P_g$ the rotor copper loss and the mechanical power developed are given by

(a) $\frac{P_g}{S}$ *and* $\frac{P_g}{1-S}$ *respectively*

(b) $S^2 P_g$ *and* $\frac{P_g}{S}$ *respectively*

(c) $SP_g$ *and* $(1\text{-}S)P_g$ *respectively*

(d) $(1\text{-}S)P_g$ *and* $SP_g$*respectively*

18. During no-load test an induction motor draws power:

(a) For core loss and copper loss

(b) For copper loss and windage and friction loss.

(c) For core loss and windage and friction loss.

(d) Only for the very small copper loss.

19. In a 3-phase induction motor, internal developed torque T in terms of supply voltage V is proportional to

(a) $\sqrt{V_1}$

(b) $V_1^2$

(c) $V_1$

(d) $\frac{1}{V_1}$

20. The torque (Nm) in an induction motor is given by the expression ($W_s$ = synchronous speed in mech rad)

(a) $\frac{3I_2^2 r_2 / S}{W_S}$

(b) $\frac{3I_2^2 r_2}{S}$

(c) $\frac{3I_2^2 r_2 / S}{(0.5P)W_S}$

(d) $\frac{3I_2^2 r_2}{W_S}$

21. The rotor of an induction motor can never run at synchronous speed, because, then the relative speed between the rotating flux and the rotor will be
    (a) Zero and hence, torque will be maximum.
    (b) Zero and hence, torque will be zero.
    (c) Maximum and hence, torque will be maximum.
    (d) Maximum and hence, torque will be zero.
22. A squirrel-cage induction motor having a rated slip of 2% on full load has a starting torque of 0.5 full-load torque. The starting current is
    (a) Twice full-load current
    (b) Equal to full-load current
    (c) Five times full-load current
    (d) Four times full-load current.
23. For induction motors, normally
    (a) The stator winding is connected to supply.
    (b) Both the stator and the rotor windings are connected to supply.
    (c) The stator winding is connected to supply and the rotor winding is short-circuited.
    (d) The rotor winding is connected to supply and stator winding is short-circuited.
24. Cogging of induction motors occurs due to
    (a) Harmonic synchronous torque only.
    (b) Harmonic induction torque only.
    (c) Vibration torques.
    (d) Both harmonic induction torques and harmonic synchronous torques.
25. The power input to an induction motor is 40 kW when it is running at 5% slip. The stator resistance and core loss are assumed negligible. The torque developed in synchronous watts is
    (a) 38 kW (b) 42 kW
    (c) 40 kW (d) 20 kW.
26. Rotor core-losses of a 3-phase induction motor are small because
    (a) Rotor flux density is very small (b) Rotor is laminated.
    (c) Rotor frequency is very small (d) None of the above.

27. In auto transformer starting of a squirrel cage induction motor the stator voltage is reduced by a factor x(< 1). Compared to direct on-line starting, the staring current and starting torque are reduced by the factors

| | Starting current | Starting torque |
|---|---|---|
| (a) | $x^2$ | $x^2$ |
| (b) | x | x |
| (c) | x | $x^2$ |
| (d) | $x^2$ | x |

28. The relationship between rotor frequency $f_2$, slip S and stator frequency $f_1$ is given by

(a) $f_2 = Sf$
(b) $f_2 = f/S$
(c) $f_2 = f\sqrt{S}$
(d) $f_2 = (1-S)f$

29. The speed of an induction motor is controlled by varying supply frequency keeping V/f constant

(a) Breakdown torque would decrease but magnetizing current would remain constant

(b) Breakdown torque and magnetizing current would both remain constant.

(c) Breakdown torque and magnetizing current would both decrease.

30. In a 3-phase induction motor, slip for maximum torque in terms of rotor resistance $r_g$ is

(a) Proportional to r
(b) Inversely proportional to r
(c) Directly proportional to $r^2$
(d) Independent of $r^2$.

31. As resistance is added in the rotor circuit of a slip ring motor

(a) Its maximum torque remains the same and also occurs at lower slip.

(b) Its maximum torque remains the same but occurs at higher slip.

(c) Its maximum torque increases but occurs at the same slip.

(d) Its maximum torque decreases but occurs at the same slip.

32. For a constant torque load voltage of a squirrel-cage induction motor is reduced by a factor of $\frac{1}{\sqrt{2}}$; its current and slip modify by factors of

| | Current | Slip |
|---|---|---|
| (a) | 2 | 2 |
| (b) | $\sqrt{2}$ | 2 |

(c) 2 √2

(d) 1/√2 √2

33. At low slip the torque-slip characteristic is

(a) $T \propto S$ (b) $T \propto S^2$

(c) $T \propto \frac{1}{S^2}$ (d) $T \propto \frac{1}{S}$

34. The numbers of poles in a 50 c/s induction motor, if the name plate speed is 460 rpm are

(a) 4 (b) 16

(c) 8 (d) 12

35. An induction motor in general is analogous to

(a) A two-winding transformer with its secondary open-circuit.

(b) A two-winding transformer with its secondary short-circuited.

(c) A three-winding transformer

(d) An auto-transformer.

36. For a slip-ring induction motor, if the rotor resistance is increased, then

(a) Starting torque increases but efficiency decreases.

(b) Starting torque decreases but efficiency increases.

(c) Starting torque and efficiency decrease

(d) Starting torque and efficiency increase.

37. Crawling of induction motors occurs due to

(a) Vibration torque.

(b) Harmonic induction torques only.

(c) Both harmonic induction torques and harmonic synchronous torque.

(d) Harmonic synchronous torques only.

38. If the mechanical load of a 3-phase induction motor is increased from no-load, then

(a) Power factor remains constant. (b) Power factor becomes zero

(c) Power factor increases. (d) Power factor decreases.

39. For a constant torque load the addition of a rotor resistance in a slip ring induction motor

(a) Causes no change in stator current but increases its slip.

(b) Reduces stator current as well as slip.

(c) Reduces stator current but increases slip.

(d) Causes no change in stator current but reduces its slip.

40. The starting current of an induction motor is five times the full-load current while the full-load slip is 4%. The ratio of starting torque to full-load torque is

(a) 0.8 (b) 1.2

(c) 1.0 (d) 0.6.

41. Maximum torque of a 3-phase induction motor is

(a) Inversely proportional to $r_2$ (b) Independent of $r_2$

(c) Directly proportional to $r_2$ (d) Proportional to $r_2^2$.

42. By adding a resistance in the rotor circuit of a slip ring induction motor

(a) The starting current and torque both increases.

(b) The starting current increases but starting torque decreases.

(c) The starting current and torque both reduce (compared to direct on-line starting).

(d) The starting current reduces but starting torque increases.

43. The stator winding of a 3-phase induction motor is connected to balanced supply and the rotor is short-circuited, then the rotor would

(a) run against the direction of rotating flux.

(b) Not run.

(c) Run synchronously in the direction of the rotating flux.

(d) Run synchronously with the rotating flux.

44. An induction motor, DC shunt motor and synchronous motor are started at reduced voltage such that the starting current is the same as the full-load current.

Mark the correct statement below in respect of starting torque

| **Induction motor** | **DC shunt motor** | **Synchronous motor** |
|---|---|---|
| (a) Full load torque | Less than full-load torque | Less than full-load torque |
| (b) Low starting torque | Much less than full-load torque | Same as full-load torque |
| (c) Less than full-load torque | Same as full-load torque | Zero starting torque |
| (d) More than full-load torque | More than full-load torque | More than full-load torque |

45. In a three-phase induction motor with increase in load from light load

(a) Stator p increases while rotor pf decreases

(b) Both stator and rotor pf decrease

(c) Both stator and rotor pf increase

(d) Stator pf decreases while rotor pf increases.

46. The relative speed between the stator and the rotor fluxes is equal to

(a) $(n_s+n)$ rpm
(b) $n_s$ rpm
(c) $(n_s-n)$ rpm
(d) zero rpm

47. Assume that the stator impedance of an induction motor is negligible. The frequency of voltage applied to the motor is increased keeping V/f constant. The maximum torque will

(a) Increase and will occur at large values of slip.

(b) Remain constant but will occur at smaller values of slip.

(c) Increase and will occur.

48. At a slip of 4%, the maximum possible speed of a 3-phase squirrel cage induction motor is

(a) 1440 r.p.m.
(b) 1500 r.p.m.
(c) 2800 r.p.m.
(d) 3000 r.p.m.

49. For a 3-phase induction motor, if the leakage reactance is reduced by using open slots, then

(a) Pull-out torque decreases.

(b) Starting current increases but starting torque decreases.

(c) Starting torque and starting current increase but power factor decreases.

(d) Starting torque and starting current decrease but power factor increases.

50. If the stator impedance is neglected, the maximum torque of an induction motor occurs at a slip of

(a) $S = \sqrt{\frac{r_2}{x_2}}$

(b) $S = \frac{r_2}{x_2}$

(c) S=1

(d) torque-slip curve does not exhibit a maximum.

51. In dynamic braking

(a) Any two stator terminals are earthed.

(b) The stator terminals are switched over to a DC source from the DC supply.

(c) The supply terminals of any two of the stator phases are interchanged.

(d) A DC voltage is injected in the rotor circuit.

52. The applications where speed ratios other than 2 : 1 are required, speed control can be done by variation of number of stator poles using

(a) Consequent pole technique.

(b) Pole amplitude modulation technique

(c) Multiple stator winding.

(d) None of the above.

53. Speed control by variation of number of stator poles is applicable to

(a) Squirrel cage motors only.

(b) Both squirrel cage and wound-rotor motors.

(c) Wound-rotor motors of large rating only.

(d) Wound-rotor motors of small rating only.

54. Star-delta starting is equivalent to auto-transformer starting with

(a) 58% tapping (b) 85% tapping

(e) 33% tapping (d) 52% tapping.

55. If the stator impedance is neglected, the maximum starting torque in an induction motor occurs at (usual symbols are used)

(a) $r_2 = 0$ (b) $r_2 = 2x2$

(c) r2 = x2 (d) r2=x2/2

56. Variation of supply frequency for speed control is generally carried out with

(a) $V_1 f$ constant (b) constant $V_1/f$

(c) constant $V_1$ (d) None

57. In stator impedance starting of a squirrel cage induction motor, the stator current is reduced by a factor x compared to direct-on-line starting. The starting torque is reduced by the factor (compared to direct-on-line starting)

(a) $x^2$ (b) 1/x

(c) $1/x^2$ (d) x

58. Regenerative braking occurs when

(a) The load is lifted by a hoisting machine.

(b) The load is lowered by a hoisting machine.

(c) The number of poles is decreased in a pole-changing motor.

59. For wound-rotor motors, rotor resistance starting is preferred over reduced voltage starting because it

(a) Limits starting current, increase starting torque and also improves starting power factor

(b) Improves starting power factor only.

(c) Increases starting torque only.

(d) Limits starting current only.

60. If stator impedance is neglected, the maximum torque in an induction motor is given by the expression

(a) $\frac{3}{w_S}\frac{V^2}{2r_2'}$ (b) $\frac{3}{w_S}\frac{V^2}{2x_2'}$

(c) $\frac{3}{w_S}\frac{V^2}{2x_2'}$ (d) $\frac{3}{w_S}\frac{V^2}{r_2'}$

61. When it is required to control the speed of the motor during deceleration, it is preferable to employ

(a) Plugging (b) Mechanical brakes

(c) DC dynamic braking (d) Regenerative braking.

62. Speed control by supply voltage variation is not done because

(a) The range of speed control is limited.

(b) It reduces pull-out torque and also the range of speed control is

l imited

(c) It reduces pull out torque only.

(d) None of the above.

63. At sub-synchronous speeds, in Scherbius system, the electrical power at slip frequency fed to the auxiliary commutator machine is

(a) Dissipated as heat.

(b) Converted to mechanical power and feel to the driven shaft.

(c) Converted to electric energy at line frequency and returned back to the supply.

(d) none of the above.

64. In an induction motor the stator m.m.f. comprises

(a) m.m.f. required to cancel rotor m.m.f.

(b) magnetizing m.m.f. only.

(c) m.m.f. to rotor m.m.f.

(d) vector sum of magnetizing m.m.f. and component to cancel rotor m.m.f. vector

65. Squirrel cage motors of large power rating are started at reduced voltage became it

(a) Increases starting torque.
(b) Increases starting power factor
(c) Reduces starting current.
(d) None of the above.

66. In applications where reversal of direction of rotation is required, use is made of

(a) AC dynamic braking
(b) Regenerative braking
(c) Plugging
(d) DC dynamic braking.

67. Direct-on-line starting is applicable to

(a) Both squirrel cage and wound rotor motors of small power capacity.
(b) Only wound-rotor motors of small power capacity.
(c) Only squirrel cage motors of small power capacity.
(d) Both squirrel cage and wound-rotor motors of large power capacity.

68. Smooth speed control can be obtained by

(a) Rotor slip-power control only.
(b) Bariation of rotor resistance.
(c) Both variation of supply frequency and rotor slip power control.
(d) Variation of supply frequency only.

69. If stator impedance is neglected, the maximum torque in an induction motor occurs at a rotor resistance of

(a) $(1\text{-}S)x_2$
(b) $x_2$
(c) $Sx_2$
(d) $(1+S)x_2$

70. Plugging is executed when

(a) The supply terminals of any two stator phases are interchanged.
(b) Two stator terminals are shorted together.
(c) Any two stator terminals are connected to a DC source.
(d) Any two stator terminals are earthed.

71. The ratio of starting torque in auto-transformer starting with x% tapping to that in direct-on-line starting is equal to:

(a) $\sqrt{x}$
(b) $x^2$
(c) x
(d) 1/x

72. At sub-synchronous speeds, in Kramer system, the electrical power at slip frequency fed to the auxiliary commutator machine is

(a) Dissipated as heat

(b) Converted to electric energy at line frequency and returned back to t the supply.

(c) Converted to mechanical power and fed to the driven shaft.

(d) None of the above.

73. For starting squirrel cage induction motors reactors are preferred over resistors because reactors

(a) Improve starting power factor.

(b) Increase starting torque

(c) Incur less power loss and effectively reduce the applied voltage to the motor.

(d) None of the above.

74. In the cumulative cascade connection of two induction motors, the synchronous speed of the set is given by

(a) $\frac{120f}{P_1}$ (b) $\frac{120f}{P_2}$

(c) $\frac{120f}{P_1 - P_2}$ (d) $\frac{120f}{P_1 + P_2}$

76. For wound rotor motors while speed is controlled by variation of rotor resistance, at low speeds

(a) Both starting torque and efficiency are decreased.

(b) Both starting torque and efficiency are increased.

(c) Starting torque is reduced and efficiency is increased.

(d) Starting torque is increased and efficiency is decreased.

77. In an induction motor the supply frequency is 50 Hz. The rotor frequency is 4.5 Hz when the slip is 7.5%. Then the rotor frequency for the slip 2.5% is

(a) 13.5 Hz (b) 0 Hz

(c) 1.5 Hz (d) 1 Hz

78. Direct-on-line starting is applied to small motors only because it

(a) Reduces starting torque.

(b) Causes objectionable voltage rise in the supply.

(c) Causes objectionable voltage drop in the supply.

(d) None of the above.

79. In the cascade connection of two induction motors, the mechanical power developed and the electrical power in the rotor of the main induction motor having pi no. of stator poles will be in the ratio of

(a) P1:P2
(b) (P1+ P2): P2
(c) P2 : P1
(d) (P1-P2): P2

80. During the blocked-rotor test on an induction motor the power is drawn mainly for

(a) Core loss
(b) Copper loss and core loss.
(c) Copper loss.
(d) Copper loss, core loss, and windage and friction loss.

81. Speeds higher than the synchronous speed can be obtained by

(a) Rotor slip power control.
(b) Variation of supply voltage.
(c) Variation of supply frequency.
(d) Variation of rotor resistance

82. The diameter of an induction motor circle diagram is (usual symbols are used)

(a) $\frac{V}{r_2}$
(b) $\frac{V}{x_2}$
(c) $\frac{V}{x_1 + x_2}$
(d) $\frac{V}{r_1 + r_2}$

V = stator phase voltage

83. The applications where speed ratios other than 2: 1 are required, speed control can be done by variation of number of stator poles using

(a) Pole amplitude modulation technique.
(b) Multiple stator winding.
(e) Consequent pole technique
(d) None of the above.

84. During short-circuit test on a slip ring induction motor

(a) The rotor is open-circuited but is free to rotate.
(b) The rotor is short-circuited but is blocked from rotation.
(c) The rotor is open-circuited but is blocked from rotation.
(d) The rotor is short-circuited but is free to rotate.

85. The secondary of a linear motor normally consists of a

(a) Istributed single-phase winding
(b) Solid conducting plate.

(e) Distributed three-phase winding

(d) Concentrated single-phase winding

86. In a 3-phase induction regulator the output line voltages are in phase with the supply line voltage in

(a) Maximum buck position only.

(b) Both maximum boost and maximum buck positions.

(c) Maximum boost position only

(d) None of the above.

87. The shunt resistance circuit component obtained by no-load test of an induction motor is representative of

(a) Core loss only.

(b) Core loss and windage and friction loss

(c) Copper loss only.

(d) Windage and friction loss only.

88. In a single-phase induction regulation maximum boost is obtained when the angle between the axes of the stator winding and the rotor winding is equal to

(a) 90 electrical degrees (b) 180 electrical degrees

(c) 45 electrical degrees (d) zero electrical degrees.

89. The blocked-rotor test of squirrel-cage induction motor determines

(a) Its equivalent shunt resistance and reactance as seen from the stator.

(b) Its equivalent series resistance and reactance as seen from the stator.

(c) Its equivalent shunt resistance and reactance as seen from the rotor.

(d) Its equivalent series resistance and reactance as seen from the rotor.

90. The synchronous impedance of a synchronous-induction motor is much larger than of a synchronous motor of the same rating because of

(a) Larger air-gap in the former

(b) Less magnetic reluctance of the former.

(c) The DC excitation to be supplied to the three-phase rotor of the former.

(d) The presence of damper bars in the former.

91. Compensating winding in a single-phase induction regulator is provided to

(a) Make the reactance of the primary winding very high at all rotor positions.

(b) Make the reactance of the primary winding negligibly small at all rotor positions.

(c) Obtain wider variation of output voltage.

(d) Increase the efficiency.

92. The rotor of a synchronous induction is delta-connected. If F is the peak m.m.f. created by a certain direct current supplied to the rotor across two slip rings, then a F is the peak m.m.f. for the same current supplied across one slip ring and the other two slip rings shorted together, where a is equal to

(a) $\frac{1}{2}$ (b) $\frac{\sqrt{3}}{2}$

(c) 1 (d) $\frac{2}{\sqrt{3}}$

93. In the circle diagram of an induction motor the vertical (parallel to V-axis) line drawn from the short-circuit point to the diameter of the circle is divided by the torque line in the ratio

(a) $\frac{r_1}{x_2'}$ (b) $\frac{r_2'}{r_1}$

(c) $\frac{x_2'}{x_1'}$ (d) —

94. Position synchro's normally have

(a) Three-phase stator winding and single-phase rotor windings.

(b) Single-phase stator and rotor windings.

(c) Three-phase stator and rotor windings.

(d) Single-phase stator winding and three-phase rotor windings.

95. A synchronous induction motor is normally fed with

(a) a.c. to the stator and d.c. to the rotor.

(b) d.c. to the stator and a.c. to the rotor

(c) Superimposed d.c. and a.c. to both the stator and the rotor.

(d) Superimposed d.c. and a.c. to the stator only.

96. In star-delta starting of a squirrel-cage induction motor compared to direct on- line starting, the starting current and starting torque are reduced by the factors

| | Starting current | Starting torque |
|---|---|---|
| (a) | $\frac{1}{\sqrt{3}}$ | $\frac{1}{3}$ |
| (b) | $\frac{1}{3}$ | $\frac{1}{3}$ |

(c) $\frac{1}{3}$ $\frac{1}{\sqrt{3}}$

(d) $\frac{1}{\sqrt{3}}$ $\frac{1}{\sqrt{3}}$

97. The rotor of a synchronous induction motor is star-connected. If F is the peak m.m.f. created by a certain direct current supplied to the rotor across two slip rings, then a F is the peak m.m.f. for the same current supplied across one slip ring and the other two shorted together, where a is equal to

(a) $\frac{\sqrt{3}}{2}$ (b) $\frac{1}{2}$

(c) $\frac{2}{\sqrt{3}}$ (d) 1

98. A synchronous induction motor

(a) Has combined characteristic of synchronous motor and induction motor at starting.

(b) Starts as a synchronous motor but runs as an induction motor.

(c) Starts as an induction motor but runs as a synchronous motor.

(d) None of the above.

99. For controlling the speed of an induction motor the frequency of the supply is increased by 10%. For maximum torque to remain constant, the supply voltage must

(a) Be decreased by 10%. (b) Be increased by 10%.

(c) Remain constant (d) Be reduced or increased by 20%.

100. When the stators of two wound-rotor induction motors are fed from the same 3-phase AC supply in parallel and the rotor windings are connected in opposition, they are called

(a) Synchronous induction motors. (b) Induction voltage regulators.

(c) Power selsyns (d) Position selsyns.

101. In capacitor motor if C1 is the capacitance required for best starting torque and C2 is the capacitance required for best running characteristic, then

(a) C1 is approximately equal to C2

(b) C1 is much smaller than C2

(c) C1 is much larger than C2.

(d) Nothing can be stated definitely regarding their relative values.

102. In a shaded-pole motor

(a) The rotor runs from the shaded portion to the unshaded portion of the pole while the flux in the former lags that in the latter.

(b) The rotor runs from the unshaded portion to the shaded portion while the flux in the former leads that in the latter.

(c) The rotor runs from the unshaded portion to the shaded portion of the pole while the flux in the former lags that in the latter.

(d) The rotor runs from the shaded portion to the unshaded portion of the pole while the flux in the former leads that in the latter.

103. In a comparative study of the torque-slip characteristic of a balanced polyphase induction motor and that of a single-phase induction motor, it is found that when slip is zero

(a) The torque has non-zero positive value on both the characteristics.

(b) The torque has a non-zero positive value on the former and has a non- zero negative value on the latter.

(c) The torque is zero on both the characteristics.

(d) The torque is zero on the former and has a non-zero negative value on the latter.

104. A 3-phase induction motor, while supplying a constant load, has the fuse of one line suddenly blown off. The motor will run as a single-phase induction motor with line current nearly increased to

(a) 3 times (b) $\sqrt{3}$ times

(c) 6 times (d) 5 times.

105. Which one of the following statements is true for a repulsion-start induction motor and a repulsion induction motor?

(a) The former runs below synchronous speed on all loads while the latter can run above synchronous speed on low loads.

(b) The former can run above synchronous speed on low loads while the latter runs below synchronous speed on all loads.

(c) Both of them run below synchronous speed on all loads.

(d) Both of them can run above synchronous speed on low loads.

106. A two-phase a.c. servomotor supplied from balanced two-phase voltages of 1 p.u. develops torque of 1 p.u. at slip = 0.5 and a torque of 1.5 p.u. at slip = 1.5. The motor has been running at a steady slip of 0.5 with 1p.u. balanced two-phase voltages, when the voltage to the control phase is suddenly reduced to zero. The servomotor will now develop torque of

(a) 0.125 p.u (b) -0.5 p.u.

(c) -0.125 p.u. (d) 0.5 p.u.

107. Which one of the following statements is true regarding the use of a centrifugal switch for change-over from repulsion to induction motor operation of a repulsion start induction motor and a repulsion-induction motor

(a) A centrifugal switch in not needed in any of them.

(b) A centrifugal switch is employed in both.

(c) A centrifugal switch is employed in the latter while not needed in the former.

(d) A centrifugal switch is employed in the former while not needed in the latter.

108. Usually the starting torques available in various split phase induction motors are in the following ascending order

(a) Resistor-split, capacitor split, resistor-reactor-split.

(b) Resistor-reactor split, resistor split, capacitor-split.

(c) Capacitor-split, resistor-reactor split, resistor-split.

(d) Resistor-split, resistor-reactor split, capacitor-split.

109. The ratio between rotor input, rotor output and rotor copper losses is

(a) 1: (1-S): S (b) 1: (S-1): S

(c) S: (1-S):1 (d) 1: (S-S):1

110. In an induction motor if the ratio of the motor output to rotor input is 0.96, then the percent slip is

(a) 0.04% (b) 4.17%

(c) 9.6% (d) 4%

## Answers

| | | | | | |
|---|---|---|---|---|---|
| 1 | C | 41 | B | 81 | A |
| 2 | B | 42 | D | 82 | C |
| 3 | C | 43 | C | 83 | A |
| 4 | D | 44 | C | 84 | B |
| 5 | B | 45 | A | 85 | B |
| 6 | B | 46 | D | 86 | B |
| 7 | C | 47 | B | 87 | B |
| 8 | B | 48 | C | 88 | D |
| 9 | B | 49 | C | 89 | B |
| 10 | B | 50 | B | 90 | B |
| 11 | C | 51 | B | 91 | B |
| 12 | D | 52 | B | 92 | B |
| 13 | C | 53 | A | 93 | B |
| 14 | B | 54 | A | 94 | A |
| 15 | D | 55 | C | 95 | A |
| 16 | D | 56 | B | 96 | B |
| 17 | C | 57 | A | 97 | C |
| 18 | C | 58 | B | 98 | C |
| 19 | B | 59 | A | 99 | B |
| 20 | A | 60 | B | 100 | C |
| 21 | B | 61 | C | 101 | C |
| 22 | C | 62 | B | 102 | B |
| 23 | C | 63 | C | 103 | D |
| 24 | A | 64 | B | 104 | B |
| 25 | C | 65 | C | 105 | A |
| 26 | C | 66 | C | 106 | C |
| 27 | A | 67 | A | 107 | D |
| 28 | A | 68 | C | 108 | D |
| 29 | B | 69 | C | 109 | A |
| 30 | C | 70 | A | 110 | D |
| 31 | B | 71 | B | | |
| 32 | B | 72 | C | | |
| 33 | A | 73 | C | | |
| 34 | D | 74 | D | | |
| 35 | B | 75 | C | | |
| 36 | A | 76 | D | | |
| 37 | C | 77 | C | | |
| 38 | C | 78 | C | | |
| 39 | C | 79 | A | | |
| 40 | C | 80 | C | | |

# APPENDIX B

## Interview Based Questions Alongwith Answers

1. Transformers

1) What name is given to the coil through which current from the source flows?

Ans: The primary winding.

2) What name is given to the coil in which a current is induced?

Ans: The secondary winding.

3) What is the objection if the core is non-continuous?

Ans: With this type of core, the flux emanating from the north pole of the bar has to return to the south pole through the surrounding air, and as the reluctance of air is much greater than that of iron, the magnetism will be weak.

4) How is this overcome?

Ans: By the use of a continuous core.

5) Is this the best arrangement, and why?

Ans: No. If the windings were put on two limbs of a square continuous core the leakage of magnetic lines of force would be excessive. In such a case the lines which leak through air have no effect upon the secondary winding, and are therefore wasted.

6) How is the magnetic leakage reduced to a minimum in commercial transformers?

Ans: In these, and even in ordinary induction coils the magnetic leakage is reduced to the lowest possible amount by arranging the coils one within the other, obviously well insulated from each other.

7) What is the voltage current relations in a transformer?

Ans: Primary voltage: Secondary voltage Primary turns: Secondary turns. Primary current: Secondary current = Secondary turns: Primary turns.

8) Are the above relations strictly true, and why?

Ans: No, they are only approximate, because of transformer losses.

9) Discuss about the classification of transformers

Ans: As in the case of motors, the great variety of transformers makes it necessary that a classification, to be comprehensive, must be made from several points of view, as:

1) With respect to the transformation, as

a) Step up transformers;

b) Step down transformers.

2) With respect to the arrangement of the coils and magnetic circuit, as

   a) Core transformers

   b) Shell transformers,

   c) Combined core and shell transformers.

3) With respect to the method employed in cooling, as

   a) Dry transformers; natural draught, forced

   b) Air cooled transformers draught; or air blast,

   c) Oil cooled transformers;

   d) Water cooled transformers.

4) With respect to the nature of their output, as

   a) Constant pressure transformers;

   b) Constant current transformers;

   c) Current transformers;

   d) Auto-transformers.

5) With respect to the kind of service, as

   a) Distributing

   b) Power

6) With respect to the circuit connection that the transformer is constructed for, as

   a) Series transformers

   b) Shunt transformers

10) Is it necessary to use a polyphase transformer to transform a polyphase current?

    Ans: No, a separate single-phase transformer may be used for each phase. Also, V-connection and Scott connections may be used.

11) Is there any choice between a polyphase transformer and separate single-phase transformer for transforming a polyphase current?

    Ans: Yes, the polyphase transformer is preferablc, bccause less iron is required than that would be with the several single-phase transformers. The polyphase transformer therefore is somewhat lighter and also more efficient.

12) Name two varieties of polyphase transformers.

    Ans: The core, and the shell type.

13) Discuss about the losses in the transformer.

Ans: The commercial transformer is not a perfect converter of energy, that is, the input, or watts applied to the primary circuit is always more than the output or watts delivered from the secondary winding. This is due to the various losses which take place, and the difference between the input and output is equal to the sum of these losses. They are divided into two classes:

a) The iron or core losses;

b) The copper losses.

The iron or core losses are due to

1. Hysteresis;

2. Eddy current;

3. Magnetic leakage (negligibly small).

Those which are classed as copper losses are due to

1. Heating the conductors (the 1’R loss).

2. Eddy currents in conductors.

14) What is Hysteresis Loss in transformer?

Ans: In the operation of a transformer the alternating current causes the core to undergo rapid reversals of magnetism. This requires an expenditure of energy which is converted into heat. This loss of energy is due to the work required to change the position of the molecules of the iron, in reversing the magnetization. Extra power then must be taken from the line to make up for this loss, thus reducing the efficiency of the transformer.

15) Upon what does the hysteresis loss depend?

Ans: Upon the quality of the iron in the core, the magnetic density at which it is worked and the frequency.

16) With a given quality of iron how does the hysteresis loss vary?

Ans: It varies as the 1.6 power of the voltage with constant frequency

17) In construction, what is done to obtain minimum hysteresis loss?

Ans: The softest iron obtainable is used for the core, and a low degree of magnetization is employed.

18) What is Eddy current loss?

Ans: The iron core of a transformer acts as a closed conductor in which small voltages of different values are induced in different parts by the alternating field, giving rise to eddy currents. Energy is thus consumed by these currents which is wasted in heating the iron, thus reducing the efficiency of the transformer.

19) How is this loss reduced to a minimum?

Ans: By the usual method of laminating the core.

20) In practice, upon what does the thickness of the laminae or stampings depend?

Ans: Upon the frequency.

21) Does a transformer take any current when the secondary circuit is open?

Ans: Yes, a “no load” current passes through the primary.

22) Why?

Ans: The energy thus supplied balances the core losses.

23) Between the iron or copper losses which is more important, and why?

Ans: The iron losses, because these are going on as long as the primary pressure is maintained, and the copper losses take place only while energy is being delivered from the secondary.

24) How may the iron losses be reduced to a minimum in transformer?

Ans: By having short magnetic paths of large area and using iron or steel of high permeability.

25) Define the copper losses in a transformer.

Ans: The copper losses are the sum of the $1^2R$ losses of both the primary and secondary windings, and the eddy current loss in the conductors.

26) Is the eddy current loss in the transformer conductors large?

Ans: No, it is very small and may be disregarded, so that the sum of the I2R losses of primary and secondary can be taken as the total copper loss for practical purposes.

27) What effect has the power factor on the copper losses of a transformer?

Ans: Since the copper loss depends upon the current in the primary and secondary windings, it requires a larger current when the power factor is low than when high, hence the copper losses increase with a lowering of the power factor.

28) What effect other than heating has resistance in the windings?

Ans: It causes poor regulation.

29) What is the behavior of a transformer with respect to heating when operated continuously at full load?

Ans: The temperature gradually rises until at the end of some hours it becomes constant.

30) Why is a high rise of temperature objectionable in transformers?

Ans: It causes rapid deterioration of the insulation, increased hysteresis losses, and greater fire risk.

31) How are the coils best adopted to air cooling for transformers?

Ans: They are built up high and thin, and assembled with spaces between them, for circulation of the air.

32) What are the requirements with respect to the air supply in forced draught transformers?

Ans: Air blast transformers require a large volume of air at a comparatively low pressure. The larger transformers require greater pressure to overcome the resistance of longer air ducts.

33) Explain in detail the circulation of the oil in transformer.

Ans: The oil, heated by contact with the exposed surfaces of the core and coils, rises to the surface, flows outward and descends along the sides of the transformer case, from the outer surface of which the heat is radiated into the air.

34) How may the efficiency of this method of cooling be increased?

Ans: By providing the case with external ribs or fins, so as to increase the external cooling surface.

35) In what types of transformers is this mode of oil cooling used?

Ans: Generally, in lighting transformers.

36) In what other capacities except that of cooling agent, does the oil act?

Ans: It is a good insulator, preserves the insulation from oxidation, increasing the breakdown resistance of the insulation, and generally restores the insulation in case of puncture.

37) What is the special objection to oil?

Ans: Danger of fire.

38) What kind of oil is used in transformers?

Ans: Mineral oil.

39) What are the requirements of a good grade of transformer oil?

Ans: It should show very little evaporation even at 100°C and should not give off gases at such a rate as to produce an explosive mixture with the air. It should not contain moisture, acid, alkali or Sulphur compounds.

40) What is the 'major' insulation in transformer?

Ans: The insulation placed between the core and secondary (low pressure) coils, and between the primary and secondary coils.

41) Describe the 'minor' insulation.

Ans: It is the insulation placed between adjacent turns of the coils.

Note: [Commercial or all-day efficiency is a most important point in a good transformer. The principal factor in securing a high all-day efficiency is to keep the core loss as low as possible. The core loss is constant-it continues while current is supplied to the primary, while copper loss takes place only when the secondary is delivering energy. In general, if a transformer is to be operated at light loads the greater part of the day, it is much more economical to use one designed for a small iron loss than for a small full load copper lose.]

42) What is the most efficient insulating material for transformers?

Ans: Mica

43) What is the efficiency of transformer?

Ans: The efficiency of transformers is the ratio of the electric power delivered at the secondary terminals to the electric power absorbed at the primary terminals.

44) What is the All-Day Efficiency?

Ans: This denotes the ratio of the total watt hour output of a transformer to the total watt hour input taken over a working day.

45) What are auto transformers?

Ans: In this class of transformer, there is only one winding which serves for both primary and secondary. On account of its simplicity, it is made cheaply.

46) What is regulation of transformer?

Ans: This term applies to the means adopted either to obtain constancy of voltage or current. In the transformer, regulation is inherent, that is, the apparatus automatically effects its own regulation. The regulation of a transformer means, the change of voltage due to change of load on the secondary; it may be defined more precisely as: the percentage increase in the secondary voltage as the load is decreased from its normal value to zero. The regulation is said to be "good" or "close", when this change is small.

47) What are the two principal precautions which must be observed in paralleling transformer terminals?

Ans: The terminals must have the same polarity at a given instant, and the transformers should have practically identical characteristics.

48) What may be said with respect to operating transformer secondaries in parallel?

Ans: It is seldom advantageous. Occasionally it may be necessary as a temporary expedient, but where the load is such as to require a greater capacity than that of a transformer already installed, it is much better to replace it by a large transformer than to supplement it by an additional transformer of its own size.

49) How are the secondaries arranged in modern transformers and why?

Ans: The secondary windings are divided into at least two sections so that they may be connected either in series or parallel.

50) What precaution should be taken in connecting secondary sections in parallel in core type if the two sections be wound on different limbs of the core?

Ans: It will be advisable to make the connections ample and permanent, so that there will not be any liability to a difference between the current flowing in one secondary winding and that flowing through the other.

51) What advantage has the star connection over the delta connection?

Ans: Each star transformer is wound for only 58% of the line voltage. In high voltage transmission, this admits of much smaller transformer cores than possible with the delta connection.

52) What advantages are obtained with the delta connection?

Ans: When three transformers are delta connected, one may be removed without interrupting the performance of the circuit, the two remaining transformers in a manner acting in series to carry the load of the missing transformer.

53) What kinds of transformers are used for three phase current?

Ans: Either a three-phase transformer, or a separate single-phase transformer for each phase.

54) What points are to be considered in choosing between three phase and single-phase transformers for three phase current transformation?

Ans: No specific rule can be given regarding the selection of single phase or three phase transformers since both designs are equally reliable, local conditions will generally determine which type is preferable.

55) What is the character of the construction of three phase transformers?

Ans: The three-phase transformer is practically similar to that of the single phase, except that somewhat heavier and larger parts are required for the core structure.

56) How are transformers connected for four wire three phase distribution?

Ans: When the secondaries of three transformers are star connected, a fourth wire may be run from the neutral point, thus obtaining the four-wire system.

57) What is the use of 'Scott connection'?

Ans: In railway AC power distribution from three phase grid power to two phase traction power-one phase in up line and another phase in down line. Also this connection is utilized in operation of single phase furnaces from three phase bulk supply.

58) Where core-less transformer can be used?

Ans: In induction furnaces.

59) What is the type of core in transformer?

Ans: Either core type or shell type and made of CRGO (Cold Rolled Grain Oriented) steel, laminations.

60) What is the difference between power transformer and distribution transformer?

Ans: The efficiency of the power transformer is nearly constant throughout the day (24 hours), while that of distribution transformer varies during different periods of the day. Power transformers are almost at rated loading throughout the day, on the other hand distribution transformer loading increases during evening periods.

61) If DC is applied in transformer primary, what will happen?

Ans: As there is no inductance offered by the winding to the applied de voltage, heavy current will pass through the primary which may seriously damage transformer winding. Moreover, in DC there is no question of induction and there will be no induced voltage in secondary.

62) If an ac voltage of lower frequency is applied at transformer primary rated for 50 Hz., what will happen?

Ans: Eddy current loss and hysteresis loss will reduce i.e., core loss will decrease.

63) Why the power rating of a transformer is usually expressed in KVA?

Ans: If the power rating is expressed in KVA, it becomes independent of variation of power factor and hence the representation of power rating in KVA is more reasonable.

64) Why are the tap-changers provided in power transformers?

Ans: To take care of the voltage drops, distribution transformers up to 1500 KVA, 11000/420V, are provided with tapings from the winding to brush studs, arranged in a circular form, voltage drops are ± 2 %, and ± 5% on the H.V. side.

## 2. DC Machines

1) How should a DC shunt generator be started?

Ans: Usually the machine is brought up to rated speed by the driving machine (engine or turbine).

2) How should a series generator be started?

Ans: The external circuit must be closed and the machine should be brought up at rated speed.

3) What do you mean by "Voltage Build-up"?

Ans: It means gradual voltage rise to maximum during starting.

4) At what point on the commutator should the brushes contact?

Ans: The brushes should be at opposite extremities of the diameter of the armature on the commutator.

5) Should the brushes be placed in the neutral plane?

Ans: No, the brushes must be advanced beyond the neutral plane to avoid sparking due to armature reaction.

6) How the number of parallel paths can be increased?

Ans: By increasing the number of poles.

7) State, what is the type of current path through the armature?

Ans: The current path is parallel in case of lap winding and series parallel in case of wave winding.

8) In the operation of DC generator, how do the poles induce in the armature affect the field flux?

Ans: It distorts the field flux into an oblique direction (armature reaction).

9) Why does this reaction require more power to drive the machine?

Ans: As the direction of the induced current and hence flux, opposes the field flux, it requires more power to keep the level of original flux.

10) What are the remedies of armature reaction?

Ans: (a) Lengthening of the air-gap (b) By making longer pole piece in the trailing side and shorter in the advancing side of the pole (c) By slotting the pole pieces (d) By adjusting the brush positions (e) By using auxiliary poles.

11) What is "angle of lead"?

Ans: It is the angle between the geometrical neutral plane and magnetic neutral plane.

12) How sparkless commutation can be achieved?

Ans: By varying the angle of lead with change of load.

13) How the eddy current loss in a DC machine can be reduced?

Ans: By laminating the armature particularly. It should be laminated at right angles to its axis.

14) Why slotted armature is preferred?

Ans: As soon as a current is induced in an armature coil by moving it in the magnetic field, the direction in the induced current is such as to oppose the motion producing it. This is called magnetic drag. In slotted armature most of the flux passes through the teeth of the armature without having a uniform distribution around it. The lines of force under one pole tend to crowd towards the nearest teeth and are less dense in opposite slots. This improves better armature reaction due to increased flux density through the teeth. Moreover, ventilation is better and effective reluctance of the air gap is reduced. Magnetic drag on coils in the slots is less while the drag is only evident in teeth. In slotted armature the coil sides are held firmly in place and the winding protection is better; moreover, due to flux concentration at the teeth, eddy current losses in the armature conductors are lesser. Hence slotted armatures are preferred in comparison to smooth embedded armatures.

15) What are the armature losses in a DC machine?

Ans: (a) Electrical losses comprising of copper loss, eddy current loss and hysteresis loss. However, hysteresis and eddy current losses are much smaller in comparison to copper loss.

(b) Mechanical losses comprising of frictional losses of bearings and brushes. Also aerodynamic losses due to air drag on armature.

16) How the core loss or iron loss of the armature can be reduced?

Ans: The eddy current and hysteresis loss in the armature has been termed as core loss and iron loss. The eddy current loss diminishes with lamination of iron parts in armature. The hysteresis loss can be reduced only by the use of low hysteresis materials.

17) Does commutation depend on brush?

Ans: Yes, upon the width of the brush.

18) What is the difference between magnetic neutral plane and geometrical plane?

Ans: The magnetic neutral plane is the position of the zero induction with distorted field flux, armature reaction being present. The geometric neutral plane is the position of the zero induction assuming no distortion of the field.

19) What is the plane of maximum induction?

Ans: In a by-polar machine, with no flux distortion considered, it is the plane 90% in advance to the geometrical neutral plane.

20) What is the effect of field distortion on commutation?

    Ans: The geometrical neutral plane does not coincide with magnetic neutral plane.

21) What is the cause of sparking at the brushes?

    Ans: It is for the self-induction of voltage in the coil undergoing commutation.

22) How this sparking can be reduced?

    Ans: (a) By brush shifting (may not be successful for heavy machines.)

    (b) By using carbon brushes (c) By providing inter poles.

23) What is the problem of having very thin and very thick brushes?

    Ans: Time must be allowed for reversal of current during commutation. Hence if the brushes are very thin commutation may not be successful. On the other hand, if the brushes a very wide the period of commutation will be lengthened.

24) Where copper brushed may be used?

    Ans: For the DC machines that require heavy current for short duration and low voltage, copper brushes can be used (e.g., automobile dynamos).

25) What should be the minimum width of the brush contact surface?

    Ans: It may be made as one and one-half times the thickness of the commutation segment.

26) What is brush voltage drop?

    Ans: It is the potential drop due to flow of current between commutators segment and brush, it varies from 0.8 to one volt per brush.

27) How the brush pressure is secured

    Ans: By regulating the tension of the springs provided in the brush holder. The tension should be such that the contacts are just made and reliable.

28) If a shunt generator does not have built-up voltage, what are likely to be the troubles?

    Ans: (a) speed may be very low.

    (b) brushes may not be in position.

    (c) resistance of the external circuit may be too small.

    (d) loose connection within the machine.

    (e) the field might have lost residual magnetism. (The field resistance may be above the value of critical resistance.

29) When a shunt generator loses its residual magnetism, how can it be again made for voltage build up?

Ans: By temporarily magnetizing the field.

30) How does overloading affect a DC generator?

Ans: Overloading causes overheating of the armature conductors that destroys of armature and commutator.

31) What is the indication of insufficient lubrication?

Ans: Bearings will be unduly heated.

32) What is the cause of excessive terminal voltage of the shunt generator?

Ans: Due to over-excitation of the field magnet or very high speed.

33) What troubles are encountered with short-circuiting of armature and commutator?

Ans: It results in over-heating and sparking. It may even cause burning of the coil.

34) What are the causes of heating in bearings?

Ans: Overheating in bearings is caused due to following reasons: -

(a) leak of bearing oil

(b) presence of dirt in the bearing.

(c) bearings not in line

(d) bearings too tight

(e) rough shaft

(f) improper centering of armature.

35) What will happen if the commutator is overheated?

Ans: It will decompose the adjacent mica insulation and also decompose carbon brush. This results in covering of the commutator with a black film.

36) What are the causes of heating in armature?

Ans: (a) short circuit in coils

(b) eddy currents

(c) moisture in coil insulations

(d) unequal strength of magnetic poles

(e) higher voltage

(f) low speed

(g) overloading.

37) What are the causes of overheating in field magnets?

Ans: (a) eddy current in pole pieces

(b) excessive field current

(c) short circuit in field coil.

38) How the insulation of any winding is affected by moisture?

Ans: Moisture decreases the insulation resistance, thus in effect produces a short circuit.

39) Why the current in a DC motor is less during running than when stand still?

Ans: On account of rotation within a magnetic field generator action to take place and a back EMF is produced in the armature. This opposes the applied voltage and hence the current intake of the motor is the ratio of difference of these two voltages to the internal resistance of the motor. On the other hand, at stand still there is no back EMF, thus the entire supply voltage is applied on the motor armature. This causes heavy inrush of current to the armature.

40) What are the chief components of a DC Motor?

Ans: (a) magnetic field.

(b) conductors placed normal to the field.

(c) provision for motion of the conductors.

(d) device for current reversal.

41) What is the nature of back EMF in a DC motor?

Ans: It is proportional to the velocity of rotation, strength of the field and to the number as well as arrangement of the conductors on the armature.

42) How the rotation of a DC shunt motor be reversed?

Ans: By reversing either field current or armature current.

43) What will happen if both the field and armature current are reversed?

Ans: The direction of the rotation of the motor will remain unchanged.

44) What will happen if the polarities at series motor terminals are reversed?

Ans: It will not change the direction of rotation of the motor.

45) If a shunt generator is used as a shunt motor, the polarities being unchanged, what will happen?

Ans: Its direction of rotation will not change because in case of generator the current and torque both being positive, in motor current or torque both are negative. This makes the direction unchanged.

46) What is the effect of cross magnetizing armature reaction on the main field in a DC motor?

Ans: It will shift the field in a direction opposite to that of rotation.

47) What are the characteristics of DC series motor?

Ans: The field strength increases with the motor current. Also, if the supply voltage is constant, the motor, at light load, will run at a very high speed. If the motor is loaded heavily the speed will be much less.

48) What are the applications of the series motor?

Ans: These apply where starting load is very high. Hence series DC motors are used in traction, loco motion, crane, tram, etc.

49) What is the type of speed characteristic of DC shunt motor?

Ans: Speed is almost constant with varying load.

50) What should be the condition of field coils during DC shunt motor shunting?

Ans: For proper start up the field coils must be excited at their rated values.

51) What is B.H.P.?

Ans: The net horse power produced by a drive at its shaft.

52) What is the standard torque-speed requirements of DC drives?

Ans: (a) constant torque at variable speed (cranes, elevators, hoist) (b) variable torque at constant speed (drives in machine shops) (c) variable torque at variable speed (railways).

53) How the speed of the DC motor can be adjusted?

Ans: For shunt motors by armature voltage control and by inserting variable resistance in the field circuit. Tapped field can also be used. For series motor: by varying the strength of the field of a series motor or by tapped field control.

54) What is the function of interpole in DC motors?

Ans: It is required to assist in commutation i.e., to help reversing the current in each coil while short-circuited by the brush. It reduces sparking.

55) What is the nature of interpole winding?

Ans: Interpole windings are in series with the armature.

56) How long does it take to start a DC motor?

Ans: Usually from 5 to 10 seconds.

57) How does the speed of a DC motor vary with variation of field current?

Ans: The speed of the DC motor will increase if the field current is decreased and vice versa.

58) How a wide range of speed regulation can be obtained?

Ans: By combination of both armature control and field control.

59) Why starter is needed for starting of a DC motor?

Ans: At start, the back EMF being zero, the initial inrush current to the motor armature is extremely high. This may even damage the motor. To prevent this inrush, current a starter is used which basically consists of a rheostat. The resistance of rheostat is high in series with the armature during starting. As the motor gradually picks up speed, back EMF is developed. The back EMF now can oppose the supply voltage, causing the current to remain low. The starting resistance can then be removed.

60) Why under voltage relay is provided in the starter of DC motor?

Ans: In case there is a power interruption to the motor during operation, the motor will stop fast. If the voltage is then restored, the motor starting current will be extremely high. On the other hand, if arrangements are made such that the starter is automatically off with interruption of power, the motor will not get voltage if the system voltage is restored after a disruption. The motor can only be started through the starter. The device that performs this action of automatically disconnecting the armature from the supply with power failure, is called under voltage or no voltage release.

61) If ac voltage is applied at de motor terminals what will happen?

Ans: As the polarities of field and armature reverse simultaneously hence the motor will rotate; however, the inductive effect of the armature winding will come into play and the motor current will decrease and the speed of rotation will be low. If the load is not reduced. there may be heavy sparking in the commutator leading to damage of the motor.

62) For which drives are the D.C. motors very much suitable?

Ans: Although AC motors are more universally applied and robust by construction, DC motors give sleeplessly variable speed control and good pick-up in traction. In D.C. motor, both fine and coarse speed control is possible.

63) What is the nature of the current in the armature coils of a D.C. generator?

Ans: Alternating current in nature which is converted to D.C. value by the commutator.

64) What is called the "suicide connection" of the shunt field of a D.C. generator?

Ans: In a closed loop DC generator armature connection, the control is such that the shunt field gets connected to the armature, after the field has been switched off, the result in a small current in the field which will flow in the reverse direction and finally the generator voltage will be zero. The connection is self-killing for the building up of the generator voltage.

65) Why the D.C. generator has become obsolete practically?

Ans: A rectifier has efficiency of 97 to 98%, whereas a DC generator has 75 to 80%.

## 3. Alternating Current Motors

### (a) Synchronous Motors

1) What do you mean by the term "synchronous"?

Ans: The term "synchronous" means in unison, that is, in step. A synchronous machine, then, is one in which rotor field rotates in unison or in step with the stator field of the alternating current in the armature.

2) If the field strength of a synchronous motor be altered, what effect does this have on the speed, and why?

Ans: The speed does not change, because the motor has to run at the same frequency as alternator i.e., synchronism is to be maintained.

3) How does a synchronous motor adjust itself to changes of load and field strength?

Ans: By changing the phase difference between the armature current and voltage.

4) State the disadvantages of synchronous motors.

Ans: A synchronous motor requires an auxiliary power for starting, and will stop if, for any reason, the synchronism is lost. Collector rings and brushes are also required. Synchronous motors are not desirable for driving shafts in small workshops having no other power available for starting, and in cases where frequent starting, or a strong starting torque is necessary. A synchronous motor has a tendency to hunt and requires an exciting current which must be supplied from an external source. Moreover, speed control is not ordinarily possible in synchronous motors.

5) State the advantages of synchronous motors.

   Ans: The synchronous motor is desirable for large powers, where starting under load is not necessary. Its power factor may be controlled by varying the field strength. The power factor can be made unity and, further, the current can be made to lead the voltage. It is also a constant speed drive.

6) For what service are synchronous motors especially suited?

   Ans: As power factor improving device (as synchronous condenser) and in electric clocks. Also, it works as a constant speed drive.

7) How do synchronous and induction motors compare as to efficiency?

   Ans: Synchronous motors are usually more efficient.

8) What term is applied to describe the behavior of the current when hunting occurs?

   Ans: The term "surging" is given to describe the current fluctuations produced by hunting.

9) Discuss about the chief characteristics of a synchronous motor.

   Ans: Starting- The motor must be brought up to synchronous speed without load. If provided with a self-starting device, the latter must be cut out of circuit at the proper time. The starting torque of motor with self-starting device is very small.

**Running-** The motor runs at synchronous speed. The maximum torque is several times full load torque and occurs at synchronous speed.

**Stopping-** If the motor receives a sudden overload sufficient to momentarily reduce its speed, it will stop and hence is sufficient to cause a loss of synchronism.

Effect upon Circuit- In case of short circuit in the line the motor acts as a generator and thus increases the intensity of the short circuit. The motor impresses its own waveform upon the circuit. Over excitation will give to the circuit the effect of capacitance and under excitation, that of inductance.

**Power Factor-** This depends upon the field current, waveform and hunting. The power factor may be controlled by varying the field excitation.

**Necessary Auxiliary Apparatus-** Power for starting, or if self-starting, means of reducing the voltage while starting; also, field exciter, rheostat, friction clutch, main switch and excitor switch, instruments for indicating when the field current is properly adjusted.

**Adaptation-** If induction motors be connected to the same line with a synchronous motor that has a steady load, then the field of the synchronous motor can be over-excited to produce a leading current, which will counteract the effect of the lagging currents induced by the induction motors. Owing to the weak starting torque, skilled attendance is required, and the liability of the motor to stop under abnormal working conditions, the synchronous motor is not adapted to general power distribution, but rather to large units which operate under a steady load and do not require frequent starting and stopping.

## (b) Induction Motors

10) Why is a single-phase induction motor not self-starting?

    Ans: When the armature is at rest, the currents induced therein are at a maximum in a plane at right angles to the magnetic field, hence there is no initial torque to start the motor.

11) What provision is made for starting single phase induction motors?

    Ans: A scheme is used for "splitting the phase" of the single-phase current. It converts single phase current temporarily into a two-phase current, so as to obtain a rotating field. This can be maintained till the motor is brought up to speed. The phase splitting device may then be cut out and the motor can be operated with the oscillating field produced by the single-phase current.

12) Describe briefly the operation of a polyphase induction motor.

    Ans: Its operation is due to the production of a rotating magnetic field by the polyphase current furnished. This field "rotates" in space about the axis of the armature, induces currents in the latter. The reaction between these currents and the rotating field creates a torque which tends to turn the armature, whether the latter be at rest or in motion.

13) Why are induction motors called "asynchronous"?

    Ans: Because the armature does not turn in synchronism with the rotating field, or, in the case of a single-phase induction motor, the field is oscillating. (Considering the latter in the light of a rotating field).

14) How does the speed vary?

    Ans: It is slower than the "field speed," that is, than "synchronism" or the "synchronous speed."

15) What is the difference of speed called?

    Ans: The slip.

16) What is the extent of the slip in polyphase induction motor?

Ans: It varies from about 2 to 5 cent of synchronous speed depending upon the size.

17) Why are induction motors sometimes called constant speed motors?

Ans: They are erroneously and ill advisedly, yet conveniently so called by builders.

18) Why do some writers call the field magnets and armature of the induction motor the primary and secondary, respectively?

Ans: Because, in one sense, the induction motor is a species of transformer, that is, it acts in many respects like a transformer, the primary winding of which is on the field and the secondary winding on the armature.

19) What is the essential condition for the operation of an induction motor?

Ans: The armature, or part in which currents are induced, must rotate at a speed slower than that of the rotating magnetic field.

20) What is negative slip?

Ans: Slip occurring whenever motors are driven at speeds exceeding synchronous speed.

21) Why is slip necessary in the operation of an induction motor?

Ans: If the armature had no weight and there was no friction offered by the bearings and air, it would revolve in synchronism with the rotating magnetic field, that is, the slip would be zero; but since weight and friction are always present and constitute a small load, its speed of rotation will be a little less than that of the rotating magnetic field, so that induction will take place, in amount sufficient to produce a torque that will balance the load.

22) How is slip in induction motor expressed?

Ans: In terms of synchronism, that is, as a percentage of the speed of the rotating magnetic field.

23) How does the slip vary?

Ans: It varies from about 1 percent in a motor designed for very close regulation to 40 percent in one badly designed as governed by load.

24) Why is the slip of polyphase induction motor ordinarily so small?

Ans: Because of the very low resistance of the armature, very little voltage is required to produce currents therein, of sufficient strength to give the required torque. Hence, the necessary rate of cutting the magnetic lines to induce this pressure in the armature is reached with very little difference between the field speed and armature speed, that is, with very little slip.

25) How does the slip vary with the load?

Ans: The greater the load, the greater the slip.

26) What is the construction of the yoke and laminae in induction motor?

Ans: They are in every way similar to the armature frame and core construction of revolving field alternators.

27) State an objection to very high speed of the rotating field.

Ans: The more rapid the rotation of the field, the greater is the starting difficulty.

28) Besides employing a multiplicity of poles, what other means are used to reduce the speed of an induction motor (squirrel cage)?

Ans: Reducing the supply voltage or frequency. However, supply voltage variation may cause the armature current to go beyond the limits. Cascade connection may also be used.

29) What is the general character of the field winding of induction motor?

Ans: The field core slots contain a distributed winding of substantially the same character as the armature winding of a revolving field polyphase alternator.

30) Are the poles of polyphase induction motors formed in the usual way?

Ans: They are produced by properly connecting the groups of coils and not by windings concentrated at certain points on saliant or separately projecting masses of iron, as in direct current machines.

31) How are the coils in an induction motor grouped?

Ans: Three phase windings are usually Y connected, A connection is common for L.T. motors

32) What other arrangement is sometimes used?

Ans: In some cases, Y grouping is used for starting and A grouping for running.

33) Explain the method of inserting resistances in the field of a slip ring induction motor.

Ans: Variable resistances are inserted in the circuits leading to the field magnets and mechanically arranged so that the resistances are varied simultaneously for each phase in equal amounts. These starting resistances are enclosed in a box similar to a direct current motor rheostat.

34) Is this a good method?

Ans: It is more economical to insert a variable inductance in the circuit, by using an auto- transformer.

35) What is auto-transformer or compensator method of starting?

Ans: It consists of reducing the voltage at the field terminals by interposing an impedance coil across the supply circuit and feeding the motor from variable points on its windings.

36) How is the armature winding connected for slip ring motors?

Ans: It is connected in Y grouping and the free ends connected to the slip rings.

37) Why is a single-phase induction motor not self-starting?

Ans: Because the nature of the field produced by a single-phase current is oscillating and not rotating.

38) What are the auxiliary coils sometimes called?

Ans: Starting coils.

39) What is the character of the starting torque produced by splitting the phase of a single-phase induction motor?

Ans: It does not give strong starting torque.

40) How is the plain squirrel cage armature modified to enable the motor to start with a heavier load?

Ans: An automatic clutch is provided which allows the armature to turn free on the shaft until it accelerates almost to running speed.

(c) AC Commutator Motors

41) Why do the local armature currents cause sparking in commutator motors?

Ans: Because of the sudden interruption of the large volume of current, and also because the flux set up by the local currents being in opposition to the field flux, tends to weaken the field just when and where its greatest strength is required for commutation.

42) What is the strength of the local current in commutator motors?

Ans: They may be from 5 to 15 times the strength of the normal armature current.

43) Upon what does the local armature current depend in commutator motors?

Ans: Upon the number of turns of the short-circuited coils, their resistance and the frequency

44) How can the local currents be reduced to avoid heavy sparking in commutator motors?

Ans:

a) By reducing the number of turns of the short-circuited coils, that is, providing a greater number of commutator bars;

b) reducing the frequency; and

c) increasing the resistance of the short-circuited coil circuit-a, by means of high resistance connectors; or b, by using brushes of higher resistance.

45) What are high resistance connectors in commutator motors?

Ans: The connectors between the armature winding and the commutator bars.

46) Does the added resistance of preventive leads, or high resistance brushes, materially reduce the efficiency of the machine of commutator motors?

Ans: Not to any great extent, because it is very small in comparison with the resistance of the whole armature winding.

47) What is the objection to reducing the number of turns of the short-circuited coils to diminish the tendency to sparking?

Ans: The cost of the additional number of commutator bars and connectors as well as the added mechanism.

48) What effect has the inductance of the field and armature on the power factor?

Ans: It produces phase difference between the current and impressed voltage resulting in a low power factor.

49) What is the effect of this low power factor?

Ans: The regulation and efficiency of the system is impaired.

50) How does the frequency affect the power factor in a commutator motor?

Ans: Lowering the frequency tends to improve the power factor.

51) What are the characteristics of the ac series motor?

Ans: They are similar to the direct current series motor, the torque being a maximum at starting and decreasing as the speed increases.

52) For what service is the ac series motor especially suited?

Ans: On account of its powerful starting torque, it is particularly desirable for AC traction service as well as in domestic gadgets.

53) What other means are employed in modern designs to reduce sparking in repulsion motor?

Ans: Compensation and the use of a distributed field winding being most effective, high resistance connectors, high resistance brushes, etc. are also adopted.

54) What are the names of the two classes of repulsion motor?

Ans: The simple and the compensated types of repulsion motors.

55) Describe a simple repulsion motor.

Ans: It consists essentially of an armature, commutator and field magnets. The armature is wound exactly like a direct current armature, and the windings are connected to a commutator. The carbon brushes which rest on this commutator are not connected to the outside line, however, but are all connected together through heavy short-circuiting connectors. The brushes are placed about 45 or 60" from the neutral axis. The field is wound exactly like that of the usual induction motor.

56) What is the action of this type of motor at no load?

Ans: If nothing can be done to prevent it, the motor will increase in speed at no load until the armature bursts, just as it will in a series direct current motor.

57) What name may appropriately be applied to this motor?

Ans: It may be called the "repulsion induction motor", because it is constructed for repulsion start and induction running.

58) Describe a compensated repulsion motor?

Ans: In its simplest form it consists of a simple repulsion motor in which there are two independent sets of brushes, one set being short-circuited, while the other set is in series with the winding, as in the series alternating current motor.

59) What is the starting behavior of the armature of a compensated repulsion motor?

Ans: It possesses at starting most of the apparent reactance of the motor, and the effect of speed is to decrease such apparent reactance, the latter becoming zero at either positive or negative synchronism, and negative at higher speeds in either direction.

60) What is the nature of the field circuit of the compensated repulsion motor at starting?

Ans: At starting it is practically non-inductive, the effect of speed being to introduce a spurious resistance which increases directly with the speed, and becomes negative when the speed is reversed.

61) For what use is the compensated repulsion motor especially adapted?

Ans: For light railroad service in Western countries. Also, in domestic gadgets and as drives in audio instruments.

62) When employed thus what is the method of control?

Ans: A series transformer is used in the field circuit of the repulsion motor.

63) To what important use is the repulsion principle put?

Ans: It is sometimes employed for starting on single phase induction motor. In this method, after bringing the motor up to speed, the winding is then short-circuited, upon itself, and the motor then operates on the induction principle.

64) What name is given to this type of motor?

Ans: It is called the "repulsion induction motor".

65) Why are three-phase Fractional H.P. motors preferable to single phase induction motors?

Ans: 3-phase F.H.P. squirrel cage Induction motors are utilized in brake-shoe operating small drilling machines because of easy maintenance and lower cost.

66) What is 'gap' power?

Ans: The electric power transferred to the rotor by induction across the stator and rotor air gap, is the gap power.